Biotensegrity

The Structural Basis of Life

Graham Scarr CBiol, FSB, FLS, DO.

HANDSPRING PUBLISHING LIMITED
The Old Manse, Fountainhall,
Pencaitland, East Lothian
EH34 5EY, Scotland
Tel: +44 1875 341 859
Website: www.handspringpublishing.com

First published 2014 in the United Kingdom by Handspring Publishing

ISBN978-1-909141-21-6

British Library Cataloguing in Publication Data
A catalogue record for this book is available from the British Library

Library of Congress Cataloguing in Publication Data
A catalog record for this book is available from the Library of Congress

Notice
Neither the Publisher nor the Author assumes any responsibility for any loss or injury and/or damage to persons or property arising out of or relating to any use of the material contained in this book. It is the responsibility of the treating practitioner, relying on independent expertise and knowledge of the patient, to determine the best treatment and method of application for the patient.

Commissioning Editor Sarena Wolfaard
Design direction and cover design by Bruce Hogarth, kinesis-creative.com
Index by Dr Laurence Errington
Typeset by DiTech Process Solutions
Printed in the Czech Republic by Finidr Ltd.

The
Publisher's
policy is to use
paper manufactured
from sustainable forests

Contents

The concept of tensegrity as a structural design principle has been around since the middle of the twentieth century and is currently seeing a huge increase in interest. From early forays into a new form of sculpture it is now incorporated into architecture, the engineering of deployable structures in space, and is attracting the attention of biologists, clinicians and those interested in functional anatomy and movement.

Tensegrity models emulate biology in ways that were inconceivable in the past, but it has taken a while for the concept to become widely accepted because of its challenges to generally accepted wisdom (from which it has been dismissed as both 'too simple' *and* 'too complicated'), although this is perhaps understandable.

Anatomy and biomechanics have taken many centuries to accumulate a body of knowledge that is now unrivalled in any other sphere, but this progress has mostly been driven through advances in technology, and established conventions have allowed many inconsistencies to survive long past their sell by date. The relegation of fascial and other connective tissues to mere supportive roles; the tolerance of an incomplete lever system that loads destructive stresses on tissues that would be unable to withstand them; and the adherence to a Euclidean spatial system that compelled classical mechanics into taking a dominant role over biology, are significant examples; but things have moved on. These ideas were put forward at times when there was little serious alternative, and it is the associated imagery and scientific convention that have allowed them to persist into the present, but they have led biomechanics down a blind alley from which it is only just starting to recover.

As an orthopaedic surgeon in the 1970's, Stephen Levin observed things at the operating table that were at odds with conventional biomechanics theory, and he set out to find a more satisfactory explanation. Starting with the study of creatures that stretched these theories to the limit he discovered that tensegrity provided a more thorough assessment of biological mechanics at *every* size scale (Levin, 1981), and that it was also compatible with the natural laws of physics, eventually introducing the term 'biotensegrity' to distinguish it from some engineering research that is *not* constrained by these rules in the same way. Around the same time, Donald Ingber, a cell biologist investigating the effect of physical forces on cancer cells also realised that tensegrity provided a better explanation of their changing behaviours (Ingber, 1981), and through experiment, detailed much of the cell mechanics and biochemistry that underlie cell function.

Both authors have considered biotensegrity in relation to the hierarchical organisation of living systems, from molecules to the entire body, and have written much on this subject over the past thirty years. Others have also made substantial contributions but it is only recently that a more widespread interest has been shown. Whilst much of this is due to the booming internet and new research that undermines traditional views of anatomy, clinicians are starting to recognise that biotensegrity provides a better explanation for the things that they observe in practice; and the intention of this book is to explain some of the basic principles of tensegrity and re-evaluate anatomy and biomechanics in the light of these findings.

The first chapter is an introduction to the tensegrity concept with a brief history of its development and some of the characteristics that make it important to structural biology. Chapters 2 and 3 describe the underlying geometry and basic tensegrity model, while Chapter 4 explains the shortcomings of classical mechanics. The following Chapters 5 to 8 present the research and give some examples of where it improves our understanding of biological structure. Chapters 9 and 10 re-evaluate the established terminology and look at some of the nuances of the biotensegrity model, with new directions of future research. Chapters 11 and 12 then look at how biotensegrity is transforming traditional views of anatomy and consider its implications to biomechanics, robotics and therapeutics. Just as every part of a tensegrity structure has an influence on every other part, it should be kept in mind that each section relies to some extent on all the others to be fully appreciated.

This book is not an instruction manual but a response to the frequently asked question, "*what is (bio)tensegrity*"; and it is hoped that it will inspire the reader to take a deeper look at biological structure and find their own ways of applying it. It is a personal perspective that recognises that all natural forms are the result of interactions between physical forces and the fundamental laws that regulate them, and that an appreciation of these simple precepts leads to a better understanding of the human body as a functionally integrated and hierarchical unit.

Graham Scarr
March 2014

I have friends and associates who have heard my *Introducing Biotensegrity* talk five, sometimes seven or eight times and still do not "get it". Whether it's my inability to communicate or the complexity of the topic, it usually takes some time to sink in. I presented a biotensegrity talk in the UK in 2004. Afterward, Graham Scarr came up to me and showed me a rather awkward, but remarkably well thought out, tensegrity model of an arm. We had previously corresponded by email, but I had no idea how capable he was and how much he had assimilated the biotensegrity concept. Our first conversation was brief and hampered by the fact that he speaks Midlands English, and I grew up in New York City and have a tin ear for languages. It was something right out of a "Benny Hill" sketch, but it was enough to let me know that this was a man I desperately needed on my side. Graham not only had grasped the concept on first go round, but had taken the ball and run with it. He has the background in science, the hands on skill and intuitive understanding of the body as an Osteopath, a deep interest in geometry, and the technical skill of a jeweler, a side hobby of his.

Over the ensuing years we have developed a friendship and association, and we have learned from each other. Graham's emails are challenging and thought provoking. His models (much better as time went on) help to visualize biotensegrity, but also stimulate further exploration. Graham's models have become some of the best illustrative models of biotensegrity and whenever a new idea in biotensegrity emerges, he is able to develop an illustrative model that clarifies it. I travel to Europe about once a year to lecture, and Graham is always there. (I have discovered that he is a reluctant traveler, and overcame his hodophobia to pursue his interest in biotensegrity). As my ear becomes more attuned to Midlands English, our conversations become deeper and more meaningful. When he visited me at my home outside of Washington, DC, we spent a week together exploring Snelson's Needle Tower, and the dinosaurs at the Smithsonian, exchange ideas. Our relationship has continued to mature and flourish. Graham has become one of the world's leading authorities on biotensegrity (there are only a handful), with his website growing to be the finest exposition of biotensegrity on the web. And now, his book.

Biotensegrity is a new science and we are just scratching the surface. Until now, there has not been a primer that could guide those interested in understanding its origins, explain its structure, and lead to further explorations. Graham Scarr's book succeeds in doing this, and much more. He has simplified the complex without dumbing it down. It is a book for all comers, experienced in biotensegrity or not, scientist, clinician, manual therapist, or just the curious. ***Biotensegrity: The Structural Basis of Life*** is written with clarity and humor. It respects the subject and the reader. Enjoy!

Stephen M Levin
June 2014

Acknowledgements

As an osteopath interested in the structural mechanics of the human body, it had become clear that conventional ideas were never going to explain some of the things that could be observed during treatment, but there seemed to be no alternative. At the same time, a lifelong interest in natural patterns and shapes was also going nowhere because there was no overriding concept that could unite them all together. So, when Liz Davies (2004) wrote a couple of articles that linked simple geometric shapes with complex anatomical structures, and described the principles of tensegrity as a more thorough explanation of biomechanics, I was hooked.

Within a week, all that collected information on natural geometry was retrieved from its dusty shelf; and one piece in particular caught my attention. It was Donald Ingber's article in 'Scientific American' entitled *The architecture of life* (Ingber, 1998), that I had saved as something of interest but never actually read. By the end of the second week, I had also found articles by Stephen Levin and received a flyer advertising his forthcoming lecture on biotensegrity; and started making models.

Thus it is that many individuals have contributed to my own understanding of biotensegrity, and ultimately, this book. Stephen Levin, as a friend and mentor, deserves especial mention for his patience and many hours spent in discussion. In this last respect, some of the ideas presented here result from conversations and lectures held over several years but that only partially made it into print, although they are still worth crediting; and they are noted in the text as '(Levin, 2014)' and relate to the reference section as 'Levin, 2014 – personal communication'.

Levin reasoned the merits of biotensegrity from first principles and recognized the tensegrity icosahedron at the core of biological structures, and the jitterbug as the model of a dynamic energy system. The ability of tensegrity models to become stronger as they contract is then similar to the activities of the heart, bladder and uterus, for which alternative mechanical models are insufficient in explaining. The concept of interlinked structural hierarchies, and the Möbius strip and Klein bottle as pattern generators for the interweaving of fascial sheets through multiple anatomical structures, were also conceived at an early stage in his thinking. In evolutionary terms, an organism's survival depends on its ability to remain stable at every instant of its existence, and the persistence of shear stresses during development would be unsustainable and render it vulnerable to collapse; which means that a lever system that inherently generates shear stresses is an unsatisfactory model of biomechanics. In addition, the idea of agonist and antagonist muscles within a biotensegrity system was discarded early on, and now has strong support from more recent research; and Levin's descriptions of closed-chain kinematics as a *geometric system* at the heart of biotensegrity now make more sense than just their abstract mathematics.

Many thanks also go to Danièle-Claude Martin for conversations that wrestled with the difficulties inherent in describing biotensegrity; Nic Woodhead, Chris Stapleton, Andrea Rippé and Ian Schofield for the opportunity to air and consolidate my thoughts. The support of Jayne Riley in furthering my interest in functional anatomy and suggesting that the book be written in the first place, is also particularly appreciated, as is my son Jacob Scarr for manipulating some of the tensegrity models during photographic sessions.

It is also with personal thanks to Rory James for the many images that captured the essential details and ethereal qualities of the models; Brian Tietz, Daniele-Claude Martin, Donald Ingber, Marcelo Pars, Kenneth Snelson, Theo Jansen and Tom Flemons for the use of their pictures; Gerald de Jong for producing a custom image taken from his video of the bouncing sphere; Stephen Levin, Theo Jansen and Don Edwards for allowing me to take photographs of items in their own collections; Maria Gough for providing a high-resolution copy of the 1921 OBMOKhU exhibition in Moscow; and Vytas SunSpiral for assistance in obtaining the picture of the Super Ball Bot, courtesy of NASA Ames/Eric James, with research performed by Vytas SunSpiral, Adrian Agogino and George Gorospe of NASA Ames, in the Dynamic Tensegrity Robotics Lab; Jonathan Bruce of UC Santa Cruz; Drew Sabelhaus and Alice Agogino of UC Berkeley; Atil Iscen of Oregon State University; George Korbel, Sophie Milam, Kyle Morse and David Atkinson of the University of Idaho; and the model built by Ken Caluwaerts of Ghent University.

I am also grateful to Nic Hedderly for the loan of a book on early Russian constructivist art; Darren Ainsworth for information on the anatomy and biomechanics of the horses leg; Craig Nevin for the layout of Möbius strips in the thigh and leg; David Hohenschurz-Schmidt for discussions on the relevance of biotensegrity to clinical practice, and Stephen Dibnah for bringing the particular arrangement of bubbles on the surface of a stirred cup of coffee to my attention.

Finally, and perhaps most importantly, are the staff at Handspring Publishing Limited; and particular thanks go to Sarena Wolfaard, Katja Abbott, Bruce Hogarth and Andrew Stevenson for their prompt answers to all my queries and for guiding this book through the production process.

Graham Scarr

1 Tensegrity

The physical Universe is a self-regenerative process... governed by a complex code of weightless, generalised principles. R. Buckminster Fuller (1975, 220.05)

Introduction

What is it?

The term *tensegrity* is a contraction of the words 'tension' and 'integrity' and was coined by the architect Buckminster Fuller (Fuller, 1975, 700.011) but it was his student, the sculptor Kenneth Snelson (Heartney, 2009, p. 20), who created the structure that inspired the concept (Figure 1.1). A tensegrity structure is recognised by its distinct set of compression elements (struts) that appear to float within a network of tensioned cables. Snelson's original sculpture showed two X-shaped wooden struts that don't touch each other at any point, with one suspended in the air and held in place by taut nylon cables. He explained it in this way: *"The sculpture could be put into... outer space and it would [still] maintain its form. Its forces are internally locked. These mechanical forces, compression and tension or push and pull are invisible – just pure energy – in the same way that magnetic or electric fields are invisible"* (Heartney, 2009, p. 20).

Due to these 'invisible forces' the whole concept of tensegrity can seem rather esoteric; it appears to defy reason and is sometimes difficult to grasp. Tensegrity structures function the same in any position, irrespective of the direction of gravity,

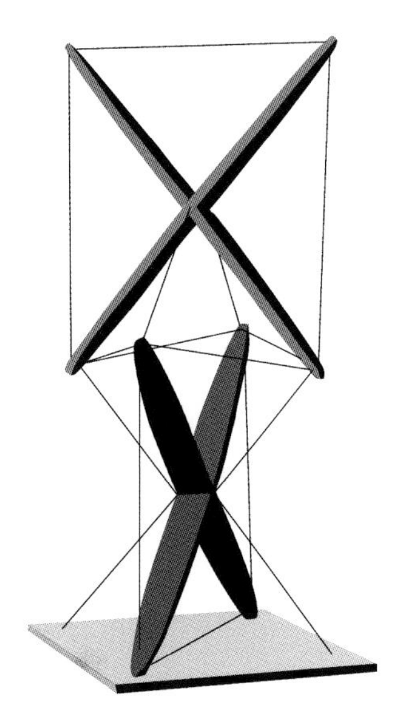

Figure 1.1
A drawing of Kenneth Snelson's *Early X-piece,* 1948

and the methods used to construct them are different from those used in traditional building. They are strong, light in weight and resilient yet can change shape with the minimum of effort and return automatically to the same position of stable equilibrium. They are also materially and energetically very efficient and have non-linear visco-elastic type properties (Juan & Tur, 2008; Skelton & de Oliveira, 2009). In fact, they are quite similar to biological structures (Gordon, 1978, p. 156).

Biotensegrity is becoming increasingly recognised as a more thorough explanation of the mechanics of motion: it examines the basic physics of natural forms through geometry and shows how even the most complicated organism can be better understood through the simplest of models. Tensegrity demonstrates the natural balance of forces, the dynamic tension network, and an integrated movement system that is applicable to all living things. Whilst the descriptions in this book include a wide diversity of organisms, the main emphasis is on human anatomy and movement. Biotensegrity is a dynamic structural system that is relevant to anatomists, clinicians and anyone interested in biology.

The origins of tensegrity

The exhibition

Karl Ioganson

The first published record of a tensegrity structure appears to be a collection of sculptures exhibited in Moscow in 1921 (Figure 1.2) by the constructivist artist Karl Ioganson (Karlis Johansons, 1890–1929) (Gough, 2005, p. 2). Constructivism, as a political and ideologically-centred art movement, favoured

Figure 1.2
A view of part of the second spring exhibition of the OBMOKhU in Moscow, 1921 (Gough, 2005, p. 76).

Figure 1.3
One of Karl Ioganson's *Spatial constructions 1920 - 1921* (Gough, 2005, p. 81).

objects that exhibited their own distinct properties and displayed how they were constructed, and its stark industrial-like designs were seen as a practice for social purpose. Some of Ioganson's sculptures were made of just three struts and nine cables, and formed structures that maintained an intrinsic equilibrium that he referred to as *Spatial constructions* (Figure 1.3). They were apparently destroyed in the 1930's although reconstructions were created for an exhibition in 1992 (Gough, 2005, p. 84). Since the early 1960's, three different people have been granted separate patents for what we now call tensegrity, which has caused a great deal of controversy and intrigue (Motro, 2003; Jauregui, 2010).

The architects

David Emmerich

David Emmerich (1925–1996), a leading architect and engineer in France, investigated structural morphology in architecture using light and auto-stable polyhedral configurations that he called "*structures tendues et autoentendants*", tensile and self-stressed structures. He received his patent in 1964 for a model that consisted of three struts connected by nine tensioned cables (Jauregui, 2010, p. 170) and was almost certainly influenced by Ioganson's designs (Motro, 2003, p. 12).

Buckminster Fuller

Prior to this, Buckminster Fuller (1895–1983), poor sighted from birth and virtually self-taught had been developing a system of geometry based on first principles. Not satisfied with conventional explanations that arbitrarily used the cube and its 90° angles as the frame of reference, he discovered that a coordinate system based on 60° and a different shape was far more fundamental (Fuller, 1975, 202.00; Edmondson, 2007, p. 124). This replaced a system that had dominated scientific measurement for generations, and as a maverick iconoclast, he refused to build on what had been done before.

Fuller looked for solutions in nature because it is structurally and functionally efficient, lightweight and dynamic, and adjusts itself to its surroundings. He perceived that all natural forms were the result of matter being acted upon by forces, and proposed them as finite energy systems consisting of the forces of tension and compression acting together. He also developed a theory that considered the shape of space as fundamental to the events of nature, and called it *synergetics*, the study of nature's coordinate system (Fuller, 1975, 200.00). By the 1940's, Fuller was using geometry to demonstrate a dynamic architecture based on nature's principles, and was spreading the word.

The sculptor

Kenneth Snelson

As a student at Black Mountain College, an experimental arts school in North Carolina, Kenneth Snelson (b. 1927) was intending to study painting. During the summer of 1948, Fuller gave a lecture on the geometry underpinning his vision of tension and compression as the most versatile forces in nature. Although not yet a celebrity, he was a mesmerising lecturer who became known as the 'Master of the Geodesic Dome', and rather fortuitously, Snelson was chosen to help with construction of the models. Snelson was so impressed that he returned a year later with his own pieces, including the simple but novel *X-piece* (Figure 1.1, page 1), and Fuller immediately recognised its significance. *"Simple as this pioneering work was, it pointed ahead to the possibility of structures in which form and function truly are...one, and the visible configuration of the sculpture is simply the revelation of otherwise invisible forces"* (Heartney, 2009, p. 20); but events took an odd turn.

Fuller recognised that this model fitted with his ideas about energetic geometry but suggested some changes. Snelson, keen to oblige his tutor, then constructed a new model that Fuller appropriated and had himself photographed alongside, but although it was displayed in numerous publications there was no mention of Snelson (Motro, 2003, p. 221). It seemed that Fuller had rationalised calling the concept his own purely on the changes he had suggested, introducing the term *tensegrity* in the 1950's, but the new model had disappeared (Jauregui, 2010, p. 27). It was not until 1959 when Fuller was exhibiting at the Museum of Modern Art, in New York, that Snelson was able to contact the curator and put the public record straight (Motro, 2003, p. 221). Even so, Fuller had already applied for a patent on tensegrity, which was granted in 1962, while Snelson's similar patent for what he called 'Continuous tension, discontinuous compression structures' was not granted until 1965. Emmerich, Fuller and Snelson had thus all received patents for what was essentially the same tensegrity structure consisting of three struts and nine cables (Jauregui, 2010, p. 165).

To be fair to Fuller, he had been working on his theory of synergetic geometry for the past twenty years (Fuller, 1975, 250.12), but Snelson's new model was now the icing on the cake. Fuller probably claimed ownership based on what he had already achieved, and even though he had inspired Snelson, it might have been

Figure 1.4
Kenneth Snelson *Superstar* 1960–2002, aluminum & stainless steel (courtesy of Stephen M. Levin)

better all-round if he had considered it a collaboration. Since then, Kenneth Snelson has forged a very successful career as a sculptor of what he calls "... *unveil[ing] the exquisite beauty of structure itself*" (Figure 1.4) (Motro, 2003, p. 225), and his pieces have been the inspiration that has taken tensegrity into the realm of biology – however, we have not finished with either of them yet.

The beginning of an idea

Tensegrity structures are stable, not because of the strength of individual members but because of the way the entire structure distributes and balances mechanical stresses (Juan & Tur, 2008). The compression elements do not touch each other at any point and are suspended by tensioned cables; the cables pull on the strut ends and try to shorten them and the struts resist this and tension the cables. The entire structure balances in a position of stable equilibrium, irrespective of the direction of gravity or a change in shape, and remains the same even if turned upside down. Tensegrity structures are light in weight and can change shape with the minimum of effort. As every component influences all the others, potentially damaging stresses are automatically distributed throughout the system so that it can react to external forces from any direction without collapsing.

Building the tradition

Traditional forms of architecture use the principle of 'blocks piled on blocks' to achieve stability and raise buildings off the ground (Figure 1.5), and bricks and stone are good at supporting compressive loads. Columns and arches keep the loads in the air by transmitting accumulated compression down to the ground, and it is the immense weight of such buildings that holds them together and enables their survival for thousands of years; but many have collapsed into piles of rubble.

Figure 1.5
Blocks piled on blocks: the traditional form of building

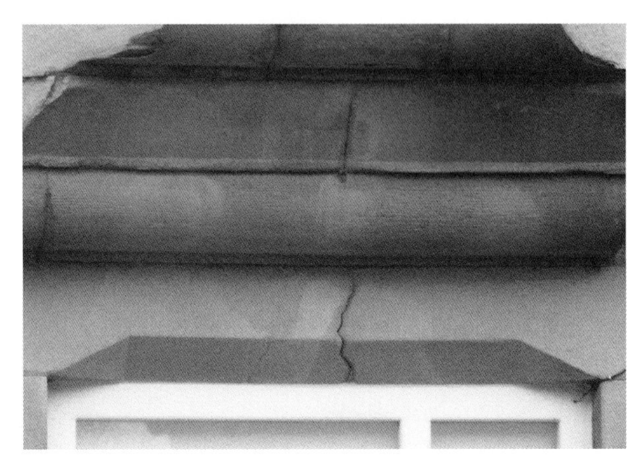

Figure 1.6
A stone lintel will crack if its lower surface is loaded under too much tension.

Stone is not very good at supporting weight if it is under too much tension and, if overloaded, the under surface of a stone lintel that is supported by columns that are too far apart will stretch and crack (Figure 1.6). The disastrous consequences of movement are clear to see after an earthquake or miscalculation in design. The addition of timber or steel beams means that potentially destructive loads can be safely carried under tension, and most buildings are now built with a rigid combination of materials loaded in tension or compression, but their weight is *always* transmitted down to the ground in a continuous transfer of accumulating compression (Figure 1.5).

A new perception

The cubic frame of reference upon which these buildings were constructed continues to dominate the mathematical system on which all other sciences are based, but this collective adherence to an outdated modality has ultimately stifled progress (Fuller, 1975, 204.00). The ancient Greeks *knew* that the earth was a sphere long before modern science 'discovered' it, and Buckminster Fuller had an inkling in the 1920's that nature could provide the principles of

Figure 1.7
An icosahedron with its faces divided into smaller triangles (1st, 2nd and 4th frequencies) takes on the appearance of a sphere.

a dynamic architecture that would improve on traditional building design. Starting from scratch, he defined a completely new approach to geometry in terms of discrete quanta, or separate bits, rather than the physically impossible continuum provided by traditional geometry (Fuller, 1975, 250.00; Edmondson, 2007, p. 162). Fuller regarded the dynamic relationship between individual components as more important than the solid, rigid and symmetrical geometry of Plato, Pythagoras and Archimedes. He used the sphere as his starting point, beginning in the centre and expanding omnidirectionally outwards, and described the connections between close-packed spheres in terms of *vectorial* relationships: force and direction; and for him the term 'frequency' denoted the various levels of complexity within the system (Figure 1.7) (Fuller, 1975, 515.00; Edmondson, 2007, p. 77). Fuller viewed planets as isolated compression elements held in place by the invisible but pervasive tension force of gravity, and realised that everything in the universe is trying to minimise energy and stabilise itself through continuous tension and local compression (Fuller, 1975, 645.00; Edmondson, 2007, p. 259).

The geodesic dome

Recognizing the sphere as the ultimate compression element because it resists compression equally from any direction (omnidirectionally), Fuller selected a shape that mimics it structurally. The icosahedron is such a shape that occupies a *greater* volume within the *smallest* surface area of any regular structure apart from a sphere; and while it only has twenty triangular faces, each one can be divided and subdivided again so that a higher frequency icosahedron gets closer to approximating the sphere (Figure 1.7), and each face is then said to be *equivalent* (Fuller, 1975, p. 58; Edmondson, 2007, p. 263).

The functional sphere

Fuller successfully established the icosahedron in popular thinking with his geodesic dome, a centerpiece of synergetic architecture seen in his most famous construction, the American Pavilion at the 1967 Montreal Expo (now known as the Montreal Biosphere): a 250-foot diameter, bubble shaped, transparent dome of steel and Plexiglas based on a high-frequency icosahedron – the world's largest dome at the time (Figure 1.8) (Jauregui, 2010, p. 1).

A geodesic dome is strong and stable yet light in weight, cheap and easy to construct, and uses essentially similar components throughout. Both

Figure 1.8
The geodesic dome designed by Buckminster Fuller for the Montreal Expo in 1967. Reproduced courtesy of © D-C. Martin: Biotensegrity, KIENER, Munich 2014.

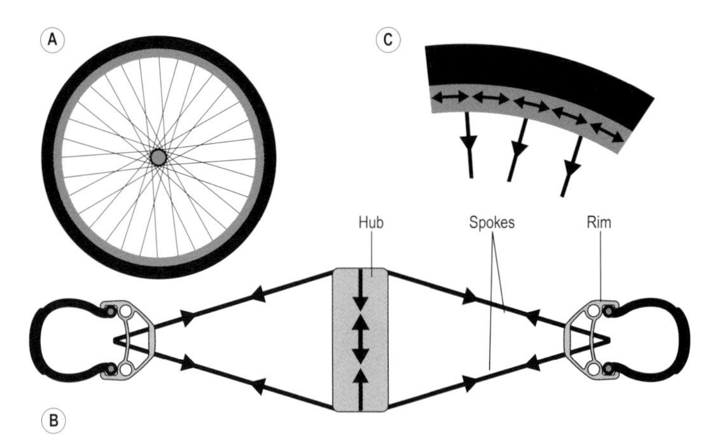

Fig. 1.9
A Axial view of a bicycle wheel; B crossed-section showing the tensioned spokes balanced by the compressed hub and rim; C close-up view of compressed rim and spokes.

internal and external stresses are distributed throughout the structure so that it is relatively immune to its surroundings, which is why it is employed in the coverings of weather stations, radar installations and satellites. The underlying geometry of geodesic domes was the inspiration that resolved the outer structure of the spherical viruses (Caspar, 1980) of which there is more in later chapters, and the third form of carbon, the fullerene molecule C60 (Kroto, 1988). Fuller reasoned that stability of the geodesic dome depended on the forces of tension and compression acting together and that this was 'built into the structure' (Fuller, 1975, 641.02). For some time before meeting Snelson, he had been searching for a model that would make the co-dependency of these two forces visible, and focused on the bicycle wheel (Figure 1.9) (Fuller, 1975, p. 354).

The bicycle wheel

Fuller noticed that the bicycle wheel differs from a cartwheel in that the central hub is suspended by the tensioned spokes attached to the outer rim of the wheel. He saw that the hub and rim were acting as isolated and *discontinuous* compression elements held together by the tensioned spokes, and that this balance maintained the wheel's integrity – the tension and compression elements remain distinct and the function of each one depends on the stability of all the others.

A combined effort

The geodesic dome and bicycle wheel were central to Fuller's ideas about tensional integrity when lecturing at Black Mountain College in 1948, but Snelson had discovered the key to suspending the compression units in a structure that had *volume*. No longer dependent on the rigidity of the geodesic dome and bicycle wheel, Fuller wrote enthusiastically to Snelson, *"If you had demonstrated this structure to an art audience it would not have rung the bell that it rang in me, who had been seeking this structure in Energetic Geometry"* R. Buckminster Fuller (Motro, 2003, p. 224).

Figure 1.10
Kenneth Snelson
Untitled Maquette,
1975, Aluminium and stainless steel; Hirschorn Museum, Washington D.C.

Both men had contributed to tensegrity but they ultimately went their separate ways. By the 1970's, Fuller was becoming a household name with his geodesic domes springing up all over the place, but he always maintained that tensegrity was really a design principle of nature (Fuller, 1975, 203.04). Snelson successfully pursued a career in sculpture but remained convinced that tensegrity had no real value except in art (Motro, 2003, p. 225). The irony of this is that his sculptures were the inspiration to others in pursuing tensegrity in biology (Figure 1.10).

Tensegrity structures are physical representations of the invisible forces within them, and the next chapter looks at how those forces interact to produce patterns and shapes that are visible in living forms. With the help of geodesic geometry, Fuller's explanations and Snelson's sculptures, the value of tensegrity as a structural and energetic system in biology will soon start to become clear.

2 Simple geometry in complex organisms

"I did not set out to design a geodesic dome. I set out to discover the principles operative in Universe. For all I knew, this could have led to a pair of flying slippers." R. Buckminster Fuller (Edmondson, 2007, p. 287)

Although Buckminster Fuller is acknowledged as a genius of the 20th century, others have considered him as a writer of little consequence. Certainly, his writings can seem rather cryptic when read line by line and should not be taken as explicitly scientific but they contain ideas that are revolutionary in their scope and depth (Edmondson, 2007, p. xxiii). Fuller looked at some of the problems facing humanity and found that much of the technology used to solve them was inadequate. He reasoned that separating technology from the natural world did not make sense, as nature had already spent billions of years achieving designs that were fluid, dynamic, lightweight and efficient in function, and thus could be very instructive. He also recognised that nature frequently displays simple patterns and shapes, and that these *do not just appear out of nowhere* but as the result of interactions between some basic rules of physics. Fuller's vision of the future was about living in harmony with nature, and he developed an entirely new approach to geometry from first principles that he called *synergetics* (Fuller, 1975, 200.00; Edmondson, 2007, p. 40).

The word *synergy* refers to the behaviour of a system that is not predicted by the sum of its individual parts but rather by their interactions, and Fuller had been working on this concept for twenty years prior to the introduction of Snelson's tensegrity model. Although the naturalist D'arcy Thompson (1860–1948) had already made a brilliant effort with his book *On Growth and Form* (Thompson, 1961), any attempt to understand the realms of greater complexity seemed impossibly complicated, and the appearance of simple geometric shapes in nature remained little more than an interesting curiosity. Fuller, however, recognised that natural structures display the inherent structural forces that are active within them, and that this was also a feature of tensegrity.

A 'new' approach to geometry: one that nature already 'knows' about

Fuller noted that everyday phenomena are conventionally described in "*... abstruse, complex [and] popularly un-teachable mathematical abstractions*" (Fuller, 1976), and that knowledge that should be experienced by everyone has been relegated to the domain of specialists. The majority of the world's population are excluded from experiencing the most efficient technology because the fundamental precepts of science and popular thinking are still based on the linear concept of a flat earth and a cubic frame of reference. After all, we still tell our children that *"the sun goes down at night"* (Edmondson, 2007, p. 4)!

Synergetic geometry regards a triangulated sphere as the most essential of structures (Fuller, 1975, 610.03) (Figure 1.7, page 5) and dispenses with all previously known standards based on the abstract mathematics of a cube. Fuller's geometry is based on a 60° coordinate system, rather than the 90° cubic system that has dominated for centuries (Fuller, 1975, 202.01), and includes motion and dynamic transformation. It is easy to model and can be demonstrated by anyone, and places the sphere at the centre of this frame of reference (Edmondson, 2007, p. 233).

We can now look at how and why natural patterns and shapes develop as they do – *first principles* – with tensegrity models emerging out of this and matching the behaviour of biology (and human movement in particular).

The rules of physics

Everything in the universe is governed by the same basic laws of physics and *always* moves towards a state of balanced equilibrium and minimal-energy. There is no question about which way is best because the choice is automatic, like an apple falling towards the centre of the earth. *Geodesic* geometry naturally follows from these rules and is the connection of any two points over the shortest distance, which is why the apple falls in a straight line, and produces the most efficient close-packing arrangements (Fuller, 1975, 410.00). Geodesic geometry makes the most economic utilization of space and materials, confers strength, and is at the heart of tensegrity and biological structure (Fuller, 1975, 700.00; Levin, 2006).

Triangulating a hexagon

We can illustrate this better by looking at a circle in 2-dimensions. A circle contains the largest area within the shortest boundary, and is a proportion that makes it a *minimal-energy* shape because it is at the very limit of this efficiency, i.e. no other shape can beat it in this respect. Circles enclose space as well as radiate outwards from their centre, as can be seen in a drop of oil floating on water, the growth of fruit mould and the ripples in a pond. Although lots of circles placed next to each other leave gaps that reduce that space-filling efficiency (Figure 2.1), joining their centres with straight lines (geodesics) reveals rigid triangles and hexagons, and makes this close-packing arrangement very strong and resilient. In fact, the most important thing to notice about geodesic geometry is that *any* three objects packed closely together will form a triangle between their centres (Fuller, 1975, 410.10).

Close-packing the shapes

Fuller considered these lines as *force vectors* that pull the centres of the circles towards each other, as well as keeping them apart. The triangles stabilise the arrangement and the hexagons represent the most energy-efficient compromise between space-filling of the circle and complete filling of the plane (Fuller, 1975, 410.00) (see footnote). Hexagons close-pack in an array that self-generates to produce the same shape at multiple size scales (Figure 2.2) and support each other where their sides meet at 3-way junctions. Even non-uniform shapes will

Figure 2.1
The close-packing of circles forms a pattern of triangulated hexagons © Rory James

approximate to hexagons when they are packed together, but they are only fully stable when triangulated (Figure 2.1).[1]

For example, bubbles spontaneously form hexagonal arrays as their surface tension reduces itself and minimises the surface area (Isenberg, 1992, p. 95) (Figure 2.3b); the arrangement of atoms in snowflakes, graphite and diamond

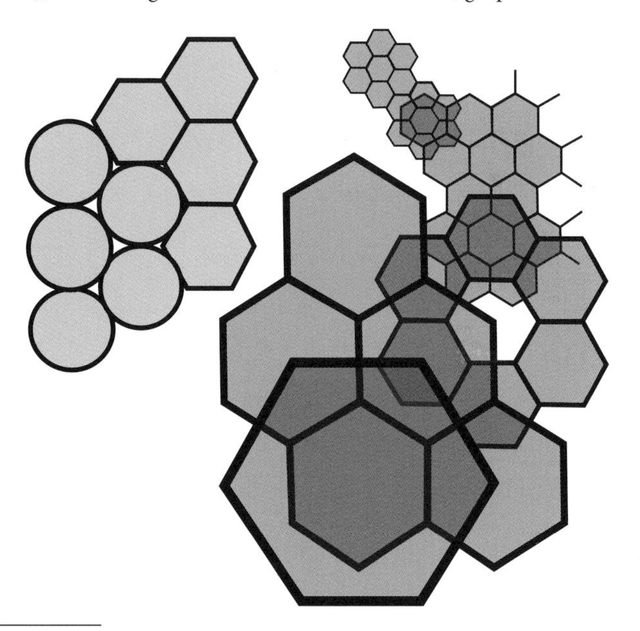

Figure 2.2
Hexagons close-pack and generate the same shape at higher scales (re-drawn from Levin, 1986).

[1] Footnote: a vector is a variable quantity that has both magnitude and direction, and the term is used here to denote that we are ultimately dealing with the forces and dynamics of natural structures.

are hexagons, as are the shapes of basalt blocks formed by the splitting of molten lava as it cools (Figure 2.3a).

These natural patterns and shapes appear because the force vectors that form them are balanced in the most energetically efficient configurations (Vanag and Epstein, 2009); and in biology, cells that are compressed by surrounding tissues also tend to approximate hexagons (Thompson, 1961, p 103). This packing arrangement is also found within myofibrils (Standring, 2005, p 118), the microstructure of the avian lung (Maina, 2007), the optic lens (Bassnett, et al., 1999) and honeycomb of the bee, etc (Figure 2.4).

Triangulated hexagons naturally appear because of the interlinked principles of geodesic geometry, close-packing and minimal-energy. In 3-dimensions these precepts lead to some new shapes that Plato described in the fourth century BCE

Figure 2.3
Hexagonal arrays: A blocks of lava on the Giants Causeway in Northern Ireland (reproduced from © Chmee2, Wikipedia); B bubbles in a glass of beer

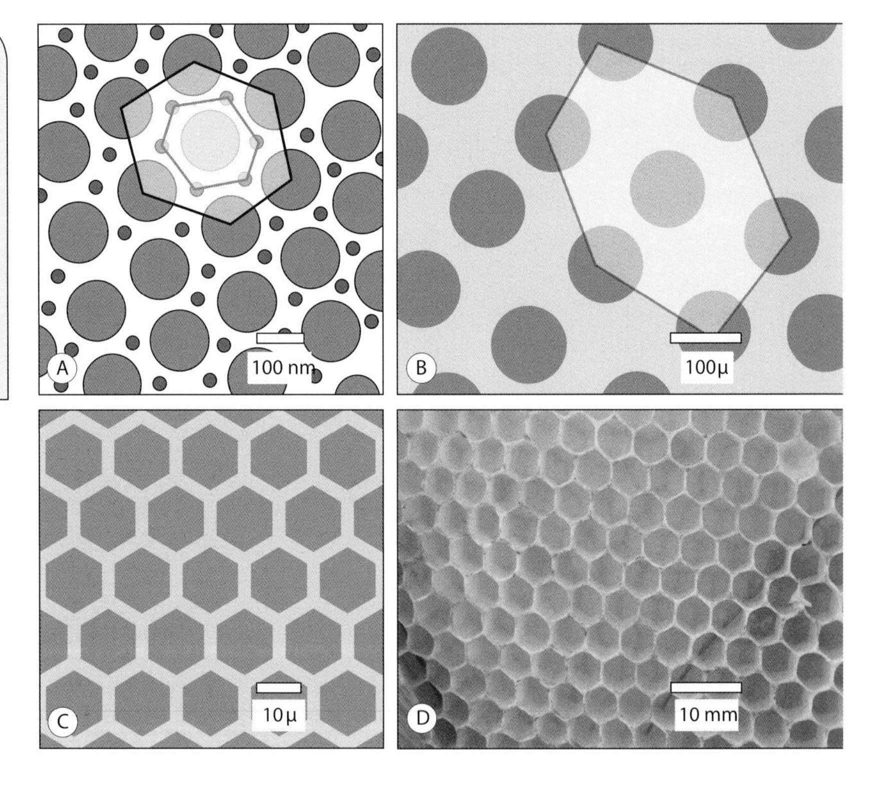

Figure 2.4
Hexagonal packing: A actin (small circles) and myosin (large circles) within a muscle fibril; B parabronchi in the avian lung (re-drawn from Maina, 2007); C epithelial fibre cells in the optic lens (re-drawn from Bassnett et al., 1999); D honeycomb.

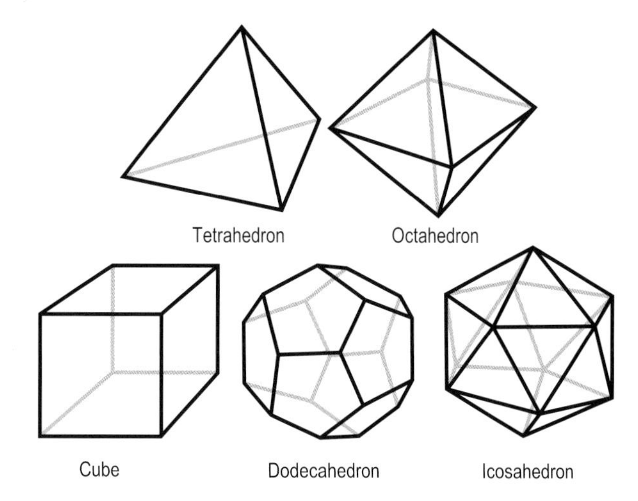

Figure 2.5
The five Platonic polyhedra (re-drawn from Scarr, 2010).

called the *Platonic solids* (Edmondson, 2007, p. 45) (Figure 2.5). Fuller recognised these shapes as important building blocks in nature because they emerge from a sphere, the centrepiece of his universal system of geometry (Fuller, 1975, 224.00). They are called *regular* polyhedra because the faces on each one are identical and meet the vertexes at the same angle; and they all have three, four or five sides.

The Platonic polyhedra

The ancient Greeks considered that these five archetypal shapes were part of natural law and described everything in the Universe because they were pure and perfect. They are particularly interesting because every vertex is equidistant from the centre, just like the surface of a sphere, and the lines between them divide the shape into equal parts (Fuller, 1975, 454.00; Edmondson, 2007, p. 232) (Figure 2.6).

A dynamic structural system

Just as the circle is the most efficient shape for enclosing space in 2-dimensions, so the sphere is its equivalent in 3-D. A sphere encloses a greater volume within the smallest surface area of *any* shape, and single atoms, bubbles, oranges and planets all approximate to spheres. Putting many spheres next to each other still leaves wasteful spaces in between, just like the circles did (Figure 2.1), but there

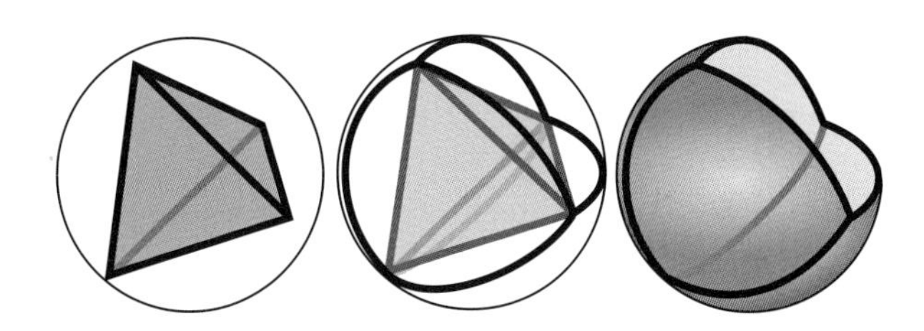

Figure 2.6
Lines connecting the vertices of the Platonic shapes form curves when projected onto the surface of a surrounding sphere, and divide it into equal parts, e.g. a tetrahedron.

is an energy-efficient solution (Edmondson, 2007, p. 115). Once again, Fuller considered the connections between the centres of the spheres to be *force vectors*, and that they represented energy and direction within what would turn out to be a dynamic system (Fuller, 1975, 223.80). In some of the models that follow, the centres of mass of the spheres correspond with the vertices of these shapes and are connected by magnetic links. The simplest of these is the tetrahedron.

The tetrahedron

Three spheres close-pack so that a line drawn between their centre points inscribes a *vector triangle* (Figure 2.1), and placing another sphere on top produces four vertices and a tetrahedron, the simplest system of *vector relationships* to have an inside and outside (Fuller, 1975, Figure 638.10; Edmondson, 2007, p. 32) (Figure 2.7). A tetrahedron occupies the *smallest* volume of unit space within the *largest* surface area, which makes it a minimal-energy shape, and naturally occurs as a pile of oranges, molecules of water and methane (Pauling, 1964), mineral crystals (Read, 1970, p. 108) and radiolaria (Haeckel, 1887, pl. 63). The addition of more spheres also connects in the same way (at 60° angles) to produce a larger second-frequency (2F) tetrahedron (see footnote) (Fuller, 1975, 225.00; Edmondson, 2007, p. 125) or a chain of tetrahedra that twists into a *tetrahelix* (Fuller, 1975, 930.00) (Figure 2.8).[2]

While the tetrahedron occupies the smallest proportion of unit space (Lord & Ranganathan, 2001a) (Figure 2.7c) the addition of more spheres in the same way forms a chain that comes close to that minimum (Figure 2.8). The *tetrahelix* is a particularly suitable model for molecular packing because of its minimal-energy efficiency, and because it introduces *chirality*, or left and right handed variations

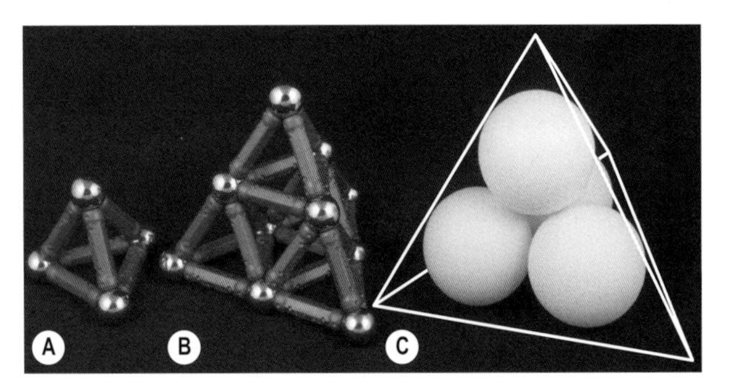

Figure 2.7
A A tetrahedron forms from the closest-packing of four spheres, and the lattice connecting the centres of those spheres represents the force vectors that link them; B the addition of more spheres to form a 2nd *frequency* tetrahedron (see footnote); C a tetrahedron occupies the *smallest* volume of space within the *largest* surface area. © Rory James

[2] Footnote: Fuller used the term *frequency* to represent expanding energy levels (the numbers refer to the connections) and the all-around (omnidirectional) growth of 'systems' (Edmondson, 2007, p. 78). It is an elegant approach to understanding the relationship between simplicity and complexity.

Figure 2.8
A chain of tetrahedra forming a right-handed tetrahelix

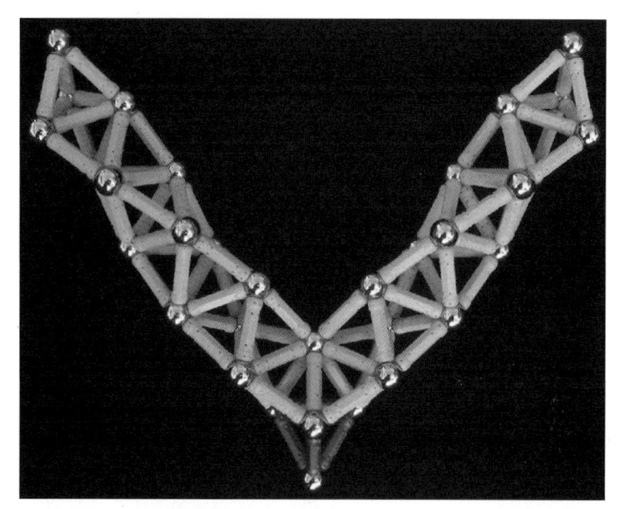

Figure 2.9
Left and right-handed tetrahelixes emerging from a tetrahedron.

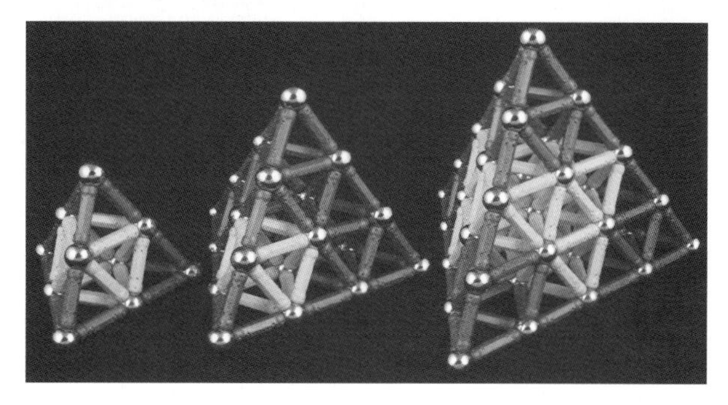

Figure 2.10
2nd, 3rd and 4th frequency tetrahedra (dark) showing the emergence of the octahedron (light), 1st and 2nd frequencies © Rory James

(Chouaib et al., 2006) (Figure 2.9); and as we shall see later, forms the geometric basis for numerous protein structures (Sadoc & Rivier, 2000; Lord, 2002): the tubular walls of plant cells (Emons & Mulder, 1998), blood vessels (Holzapfel, 2008, p. 287) and myofascia (Purslow, 2008, p. 328) – even the movement of joints is helical!

Although the tetrahedron is a minimal-energy structure, a second-frequency (2F) tetrahedron allows another shape to emerge from within this configuration: the octahedron (Figure 2.10).

The octahedron

The octahedron is formed by joining six spheres, or points, and is a shape that naturally occurs in crystals (Figure 2.11) and radiolaria (Haeckel, 1887, pl. 117).

When combined with the tetrahedron it forms the *octet truss* used by structural engineers because of its strength and stability (Fuller, 1975, 422.00; Edmondson 2007, p. 144) (Figure 2.12). Although the three central squares of the octahedron are aligned at 90° to each other, they remain stable because of the 60° cross bracing surrounding them. A good approximation of this truss appears in the bones of birds, probably due to its strength and lightness (Thompson, 1961, p. 236), but the real value of the octahedron is as part of a dynamic energy system, and we will come to that shortly.

The cube

The third of these Platonic shapes is the cube (hexahedron), and it is tempting to think that it is constructed from just eight spheres with each one occupying a corner. However, a more efficient packing arrangement has the spheres offset by 60° to each other, the same as the tetrahedron and octahedron, where *fourteen* spheres now form the unit cube (Sorrell & Sandstrom, 1973, p 121) (Figure 2.13). This cube consists of a central octahedron with a tetrahedron on each of its faces to form the corners; or alternatively, it is two 2nd frequency tetrahedra that interpenetrate each other as a duality, and because the spheres are connected at 60° and triangulated, it

Figure 2.11
A and B: 1st and 2nd frequency octahedra, © Rory James Octahedral crystals of: C iron pyrite, D diamond and E spinel.

Figure 2.12
The octet truss © Rory James

Figure 2.13
A The structure of a cube; note the central octahedron (light) and two interpenetrating 2F tetrahedra that surround it © Rory James; B twinned cubic crystals of iron pyrite with the same basic structure (courtesy of Don Edwards).

is much more stable (Fuller, 1975, p. 7; Edmondson, 2007, p. 54). Leonardo da Vinci (1452–1519) referred to it as the *star tetrahedron* (Fuller, 1975, 637.00), Johannes Kepler (1571–1630) called it the *stella octangula* (Motro, 2003, p. 99) and Buckminster Fuller named it the *duotet* (Fuller, 1975, 1033.703), so it has a fairly good provenance. The cube naturally occurs in the crystal structures of the minerals pyrite and fluorite, etc (Figure 2.13), radiolaria (Haeckel, 1887, pl. 94) and in complex organic structures such as the cores of some molecular complexes (Izard et al., 1999).

The tetrahedron, octahedron and cube are thus contained within each other and form a common lattice structure, with all the intersecting points (spheres) equidistant from each other and extending equally in all directions, and Fuller referred to it as the *isotropic vector matrix (IVM)* (Fuller, 1975, 420.00; Edmondson, 2007, p. 143).

The isotropic vector matrix and vector equilibrium

The IVM is a structural configuration that fills space completely and reveals another shape, the *cuboctahedron* – a non-regular shape that consists of twelve spheres surrounding a central nucleus (Figure 2.14). Considering this from an energy perspective, Fuller described each vertex (sphere) as the centre of a dynamic energy system, with vector lines radiating outwards and balanced by others of equal size surrounding them, and called it the *vector equilibrium (VE)* (Fuller, 1975, 430.00; Edmondson, 2007, p. 103).

Finally, the VE is an omni-symmetrical model of force vectors that extend equally in all directions and the primary coordinate system that generates all others, including tensegrity – and *bio-tensegrity is intrinsically about living geometry.* However, before we get to this we must take a brief look at the two remaining Platonic shapes, both of which are in a different structural class from the others because they only exist as a single outer shell.

The icosahedron

The icosahedron is formed by twelve spheres surrounding a central *space,* rather than the nucleus of the cuboctahedron, and is slightly smaller. While

Figure 2.14
The cuboctahedron emerging from the isotropic vector matrix (a 4th frequency tetrahedron in this example) as the closest packing of twelve spheres surrounding a central nucleus; the connections radiating from the central sphere (dark) are balanced by the circumferential ones (light). The cuboctahedron is a model of a dynamic energy system known as the vector equilibrium (VE). © Rory James

the tetrahedron contained the *smallest* volume within the *maximum* surface area of any regular shape, an icosahedron encloses the *largest* volume within the *minimum* surface area of any structure apart from a sphere, and is very stable because of its twenty triangulated faces (Figure 2.15). The icosahedron is enlarged by subdividing each of its faces and increasing the frequency (Figure 1.7, page 7), and thus comes even closer to matching the ultimate space-filling efficiency of a sphere (Edmondson, 2007, p. 264).

Natural 'spheres' are minimal-energy structures that automatically assume their shape because the pressures acting on their outer surfaces are the same in every direction (omnidirectional), and balanced, but true spheres do not exist in the real world because their smooth continuous surfaces are just mathematical constructs (Fuller, 1975, 515.01). Soap bubbles, for instance, may look like spheres at the macro-scale but when studied in detail their surfaces are really lots of irregularly shaped molecules joined together. As compliant organic structures will always try to assume the shape of a sphere (the ultimate minimal-energy structure), a high-frequency icosahedron is then the closest regular shape that they can possibly settle into (Fuller, 1975, p. 58; Edmondson, 2007, p. 263).

Some naturally occurring structures that are clearly based on the icosahedron are the molecular carbon fullerene C60 (Buckyball) (Kroto, 1988), radiolaria (Haeckel, 1887, pl. 117), 'spherical' viruses (Twarock, 2006) and clathrins (endocytic vesicles beneath the cell membrane) (Fotin et al., 2006) (Figure 2.16). Levin considered the icosahedron to be the finite-element of biological structure (Levin, 2006), and as a fundamental pattern in the structure of quasi-crystals, a relatively new field of study in the crystalline structure of metals (Lord & Ranganathan, 2001b), water (Johnstone et al., 2010) and organic

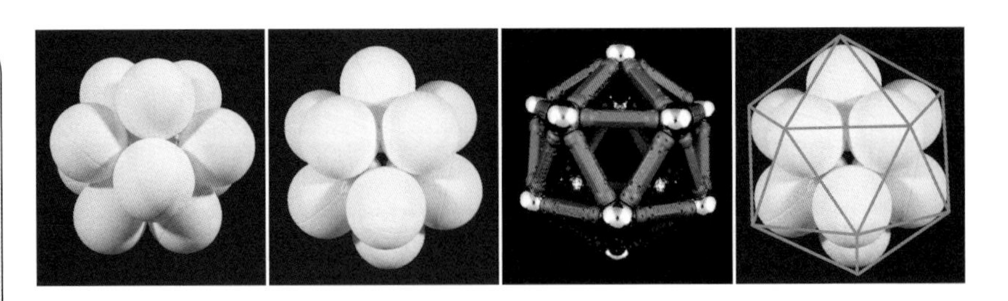

Figure 2.15
Twelve spheres close-pack around a central space to form an icosahedron that occupies the largest volume within the smallest surface area. © Rory James

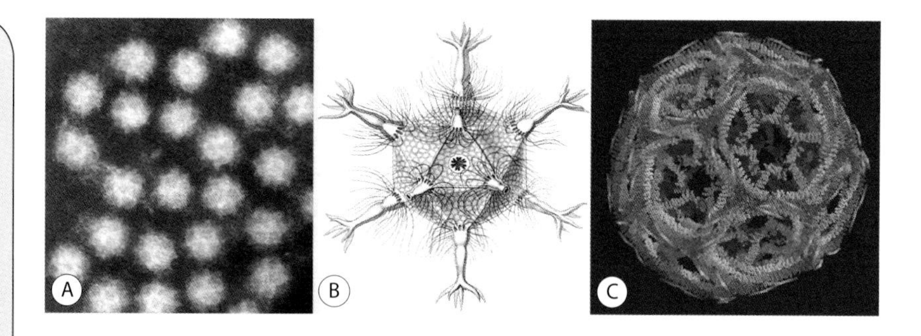

Figure 2.16
Natural structures based on the icosahedron: A sapovirus (reproduced from © Graham Colm, Wikipedia); B radiolarian (reproduced from Haeckel, 1887, pl. 117; C clathrin (reproduced from © Phoebus87, Wikipedia).

molecules (Sadoc & Rivier, 2000; Martin et al., 2000), the icosahedron is indeed worthy of more detailed consideration; and we will look at it more in the following chapters.

The dodecahedron

The last of the Platonic shapes that awaits description is the (pentagonal) dodecahedron (12 faces and 20 vertices), the dual of the icosahedron, but it cannot be constructed in the same way as the other shapes because the non-triangulated faces are unstable (Edmondson, 2007, p.48). Even so, this shape does naturally appear in pyrite crystals (Figure 2.17a), pollen grains (Figure 2.17b), the central core of some proteins (Izard et al., 1999) and in water molecules (Johnstone et al., 2010).

Having described these five fundamental shapes, we can now begin to summarise *why* they are so important to living organisms.

The geometry of living structure

Nature *always* does things in the most efficient way possible – the principle of minimal-energy – and *always* connects two points over the shortest distance – geodesic geometry. As a result, the most energy-efficient and stable way of connecting any three points together in 2-dimensions is to form a triangle, and a fourth point in 3-D, a tetrahedron (Figure 2.7). In fact, there is no other way, and although this statement might seem quite simplistic, it is a *fundamental*

Figure 2.17
Natural dodecahedrons: A iron pyrite crystal; B pollen of Morning Glory (Ipomoea purpura) x 500(reproduced from © Dartmouth electron microscope facility, Wikipedia).

physical truth and the isotropic vector matrix, tetrahelix and icosahedron are just different close-packing arrangements that result from it.

At the smallest scale, atoms relate to each other through the invisible forces of attraction and repulsion, and these forces interact in ways that *always* move the system towards a state of balanced equilibrium to form crystals (Figures 2.11 and 2.13) and molecules (Denton et al., 2002). The atoms settle into these stable and energy-efficient configurations because of the inter-related principles of geodesic geometry, close-packing and minimal-energy, i.e. the *fundamental laws of physics*, and because these are the basic rules of self-assembly, they must apply at every level in even the most complex organism (Fuller, 1975, 220.04; Levin, 2006)!

All natural structures are balanced energy systems, and it is the geodesic geometry that underlies them, and Fuller described how these energetic shapes can transform themselves from one state to another and change symmetry (Fuller, 1975, 223.80). Starting with the vector equilibrium (Figure 2.14), he showed how a simple model can demonstrate the energy characteristics of a structure that expands and contracts around a central point (and relates directly to the dynamics of tensegrity) and called it the *jitterbug* (Fuller, 1975, 460.00; Verheyen, 1989; Edmondson, 2007, p. 179) (Figures 2.18 and 2.19).

The jitterbug

From its most expanded cuboctahedral stage, the vector equilibrium twists, folds and contracts down to assume the slightly smaller icosahedral form, with the squares of the VE now contorted into rhomboids; it then continues its contraction to become an octahedron before untwisting and returning back to the original shape.

The jitterbug is not really a structure as such but an oscillating energy system in continuous motion. It contracts and expands omnidirectionally around a central point, first one way and then the other, and is a dynamic model of bodies in motion before they crystallise into form. Levin proposed the jitterbug as a model for the pumping action of the heart, bladder and uterus (Levin, 2006), and perhaps even before this, where embryonic cardiac cells rhythmically contract and start circulating blood before the heart has even formed as a distinct entity (Levin, 2014).

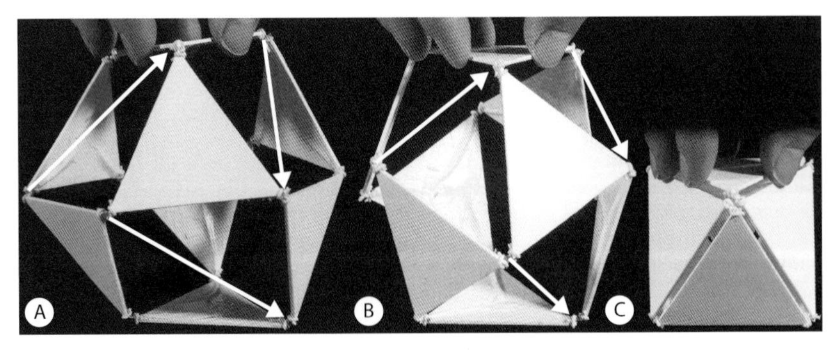

Figure 2.18
A model of the jitterbug showing: A triangles of the vector equilibrium rotating around different axes; B the system contracts down through the icosahedral stage to C, the octahedron, before returning to the original 'state'. © Rory James

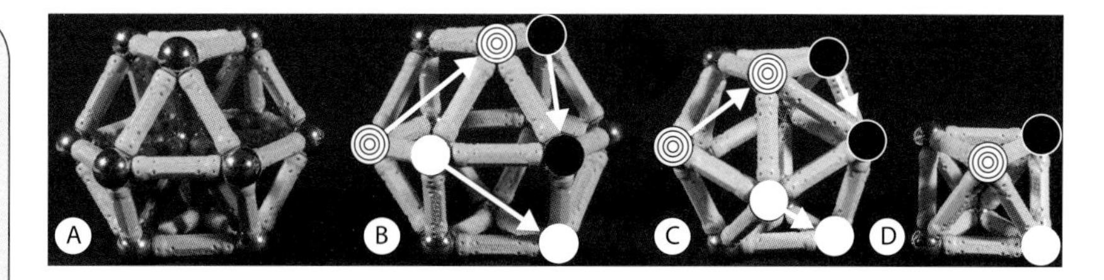

Figure 2.19
A model of the different geometric stages of the jitterbug: A cuboctahedron (for comparison); B vector equilibrium; C icosahedron; D octahedron. The arrows indicate changes within the structure as it contracts; and the white, black and lined spheres are points in this *energy system* that merge together, respectively. © Rory James

All this might seem a long-winded way of explaining biotensegrity, but it is the basis on which *all* natural structures form and their functions can be better understood. Fuller recognised that nature's design principles were involved in the transformation of polyhedral forms through the dynamic jitterbug and tensegrity concept, and was concerned with "*...mapping the intersection between the worlds of the physical and metaphysical, the oscillating continuum of symmetry and asymmetry*" Bonnie de Varco (1998, III p. 4); Snelson then displayed the beauty of structural simplicity for all to see, through tensegrity (Heartney, 2009).

3 The balance of unseen forces

A geodesic is the most economical relationship between any two events.
R. Buckminster Fuller (1975, 702.01)

Fuller stated that *"all structures, properly understood, from the solar system to the atom, are tensegrity structures"* R. Buckminster Fuller (1975, 700.04). Unfortunately, he had a habit of making grand statements that obfuscated the details of what he was trying to say, and bold though it is, this sentence has caused much confusion. If tensegrity is everywhere, what is the point of thinking about it as anything special?

We can answer this by considering the enormous range of sizes that exist within the physical world – from the level of sub-atomic particles to the entire universe, but we must consider Fuller's statement from our own viewpoint, which is somewhere in the middle. We observe the reality of solid objects but know that the atoms they are composed of are mostly empty space; we cannot *see* the forces inside them but we know they must be there. The difference between what we call 'solid' and 'empty' then comes from our own perspective, and Fuller's genius was that he understood *how* these forces were working.

All natural structures conform to the same basic rules of physics, and it is the *balance* of forces that hold them together (Fuller, 1975, 641.00). Atoms form crystals and molecules and relate to each other through the forces of *attraction* and *repulsion*, which makes them tensegrity structures in their own right (Zanotti & Guerra, 2003; Edwards et al., 2012); but that doesn't mean that the doorpost should be considered as such.

These physical principles can be illustrated with simple models that use cables and struts to represent the forces of tension (attraction) and compression (repulsion), respectively, with each component carrying just one type of load (Fuller, 1975, 720.00; Edmondson, 2007, p. 259) (Figure 3.1). They are *geodesics*

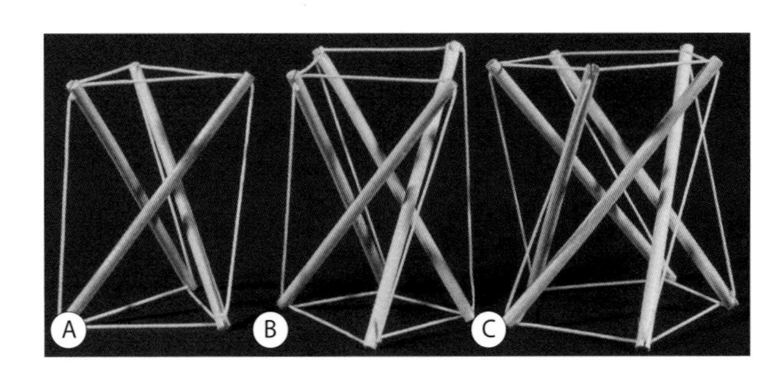

Figure 3.1
The simplest tensegrity models – all shown here with a right-handed twist (the distinction between left and right-handedness is a standard convention). © Rory James A T3-prism; B T4-prism; C T5-prism

because their structural elements are connected over the shortest distance; they are *close-packed* because they cannot get to a simpler configuration; and they automatically balance in a stable state of *minimal-energy*. Tensegrity models reduce structure to its simplest form, and the sticks-and-string are intrinsic displays of the force vectors that are active within them. An infinite number of complex configurations then become possible but they all derive from one basic model, the tensegrity prism (Pugh, 1976; Motro, 2003, p. 39).

The tensegrity model

T-prisms

The T3-prism is the simplest tensegrity model with 3-struts twisted around a central axis and a triangle of cables linking them together at each end (Figure 3.1a). It is the first of an infinite series that continues with the addition of more struts as the T4-prism, T5-prism, etc. Although they are called 'prisms', they all have a left or right-handed twist or chirality that conforms to the simple formula (Kenner, 1976, p. 8):

$$\text{Angle of twist} = 90° - 180°/n \quad \text{where } n = \text{number of struts}$$

Thus, the T3-prism has a twist of 30° between its upper and lower end, and a T4-prism has a 45° twist, etc. Interestingly, stable nano-structures of DNA based on the simple T3-prism and Platonic shapes can spontaneously assemble *in-vitro* (Liedl et al., 2010), a finding that has potential uses in nanotechnology. Fuller referred to the T3-prism as the tensegrity octahedron (six nodes/vertices) (Fuller, 1975, 724.10) and it is also known as the simplex, triplex and tripod, followed by quadruplex (T4), pentaplex (T5) and hexaplex (T6), etc; but the overall outlines of T-prisms actually match those of a class of geometric shapes known as *antiprisms*!

T-prisms have been classed as *cylindrical* tensegrities (Kenner, 1976, p. 8) because they have a single axis of rotational symmetry (similar to the struts and cables that form them). Increasing the number of struts causes them to settle more towards the outside of the cylinder and form a tubular 'wall' and when joined end to end they form a stack known as a T-helix (Figure 3.2).

T-helixes

Tensegrity helixes are not continuous, like metal springs, but form tubes with parts that are separate, just like in biology; and each cable and strut could also be made from a smaller helix, which in turn contains parts made from even smaller helixes within a structural hierarchy. Consecutive segments in the stack can

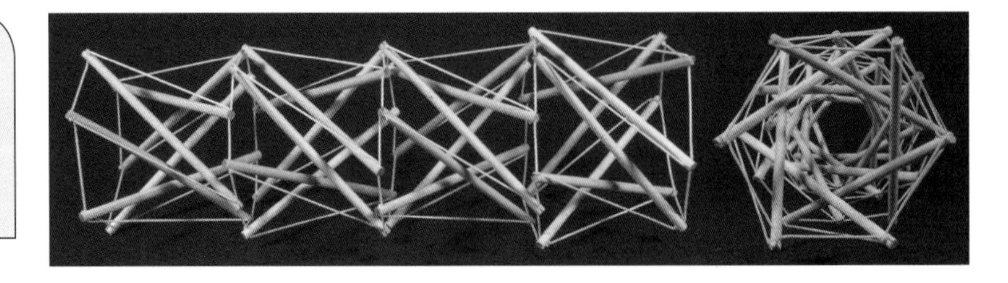

Figure 3.2
A modular chain of four right-handed T6-prisms forming a 'hollow' T-helix © Rory James

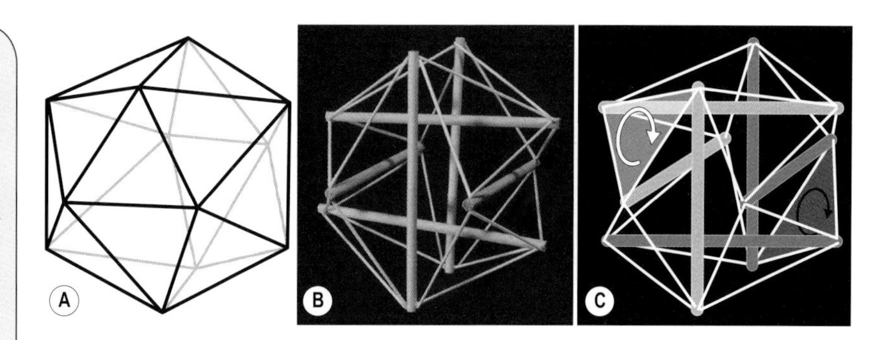

Figure 3.3
A Icosahedron; B T6-sphere or T-icosahedron © Rory James; C a group of three struts highlighted by a 'tension-triangle' form a right-handed twist (light arrow), and the opposite group with a left-handed twist (dark arrow when viewed from the other side of this model).

have the same or opposite chirality, and these influence its physical properties, so that a T-helix with segments of the same chirality will twist when stretched or compressed whereas one with alternating chiralities will resist this – a crucial property in the function of biological tubes. The geometric and structural links between the basic tetrahelix (Figure 2.8, page 17), tensegrity helixes and complex biological helixes are described in chapter 6, but in the meantime, the T3-prism has another trick up its sleeve.

The T6-sphere and tensegrity-icosahedron

The T6-sphere, or T-icosahedron (Figure 3.3), is the simplest of the *spherical* class of tensegrities (Kenner, 1976, p. 11) and is named as such because its component parts are *all* spaced equally around a central point, like an icosahedron and sphere (Fuller, 1975, 717.00; Edmondson, 2007, p. 283).

This model clearly shows three pairs of parallel struts suspended at 90° to each other, which correspond with the three-axes of 'cubic' symmetry that all the Platonic shapes have, but this observation is misleading when it comes to biology. It is much more useful to consider the six struts in different groups of three, with each strut oriented at 90° to the others *within* that group and forming a chiral twist, and struts in the opposite group having a different twist (Figure 3.3c) (Fuller, 1975, p. 400). The 'tension triangles' joining the strut ends (nodes) then highlight two opposing groups, and there are four of these groups, with each one formed from a different combination of struts. Both left- and right-handed chiralities (of what are essentially pairs of T3-prisms) are thus balanced, and link the torsion of the helix with the omni-symmetry of the sphere; and its multiple symmetries make it very useful for making even more complex biological models.

To be clear, the T6-sphere is best known as the *tensegrity-icosahedron* because the strut ends (nodes) *almost* correspond with the vertices of an icosahedron, although it is sometimes referred to as the *expanded octahedron* (Pugh, 1976, p. 26; Edmondson, 2007, p. 283) and *reducing cuboctahedron*. The reason for these different names is that this particular model corresponds closely with the icosahedral phase of the jitterbug (Figures 2.18 and 2.19, page 22 and 23), which contracts omnidirectionally *towards* the octahedral phase and expands *towards* the cuboctahedron (Figure 3.4) (Fuller 1975, 724.32), and it is partly from this perspective that Levin described its usefulness in modelling biology (Levin, 2002). Other arrangements can have nodes that *precisely* match the vertices of an icosahedron (but are structurally more complicated) and, together

Figure 3.4
Pushing opposite 'tension triangles' of a T-icosahedron towards each other causes the *whole* structure to contract around a central point, and demonstrates the oscillating features of the jitterbug (see Figs. 2.18 and 2.19, page 22 and 23). © Rory James

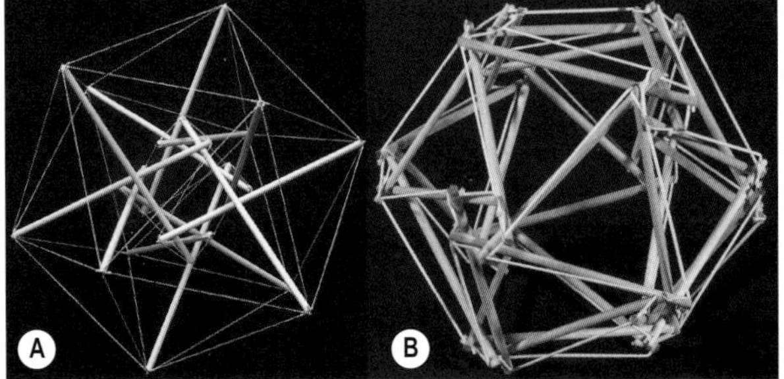

Figure 3.5
A A 12-strut tensegrity model with the nodes precisely matching the vertices of an icosahedron, reproduced courtesy of © Marcelo Pars; B a T30-icosahedron, © Rory James.

with the 30-strut model, are also referred to as the tensegrity-icosahedron (Figure 3.5) (Edmondson, 2007, p. 280; Pars, 2014). They all illustrate the variety of possibilities that are inherent in tensegrity models.

The tensegrity-icosahedron can be summed up as "*...the perfect balance between compression and tension stretched uniformly around a spherical curve, the tensegrity sphere is the truest ephemeral form visually representing the property of dynamic strength, the hallmark of all geodesics. To Fuller, it was nothing less than the perfect man-made demonstration of optimal efficiency in the use of materials*". Bonnie G. de Varco (1997, IV p. 4).

The simple complexity of tensegrity

Tensegrity models are made from just two basic components, cables and struts, and are surprisingly strong and light in weight for their size – and they become even stronger when loaded! The struts 'float' within the tensioned network of cables and are coupled into a mechanical unit that can change shape with the minimum of effort and automatically return to its original form. Tensioned cables *always* try to reduce their length to a minimum (like an elastic band), and the rigid struts limit how much they can do this, and it is the interactions between these two factors (shortening and its constraint) that allows a tensegrity system to balance itself in the most stable position of equilibrium (minimal-energy).

Essentially, each cable is under tension and every strut under compression, and because these forces always act in straight lines (geodesics), the structural elements can be optimised to their particular function; which is why they are all so long and thin. Tensegrity reduces structure to its simplest form, and the separation of tension and compression into different components means that these forces become visible through the cables and struts, because ultimately, *tensegrity structures are the physical representations of the invisible forces that hold them together.*

Structure and energy

This intrinsic link between structure and energy is an important characteristic of tensegrity and, because biology is influenced by the same rules of construction (geodesic geometry, close-packing and minimal-energy), it has been reasoned that it must also apply to living organisms (Robbie, 1977; Levin, 1981) Although it is easy to think of bones as compression struts and muscles as tensioned cables, and most of these stick-and-string models are built on a similar size scale, the essence of bio-tensegrity is structural and functional interdependency between components at *multiple* size scales. The efficient separation of tension and compression into different elements then means that the structure of each one only needs to be able to support its own particular type of load. It can also be made even more materially efficient by forming a structural hierarchy, where each cable and strut is itself made from a chain of smaller tensegrities, such as T-helixes or T-icosahedra (Figures 3.6 and 3.7).

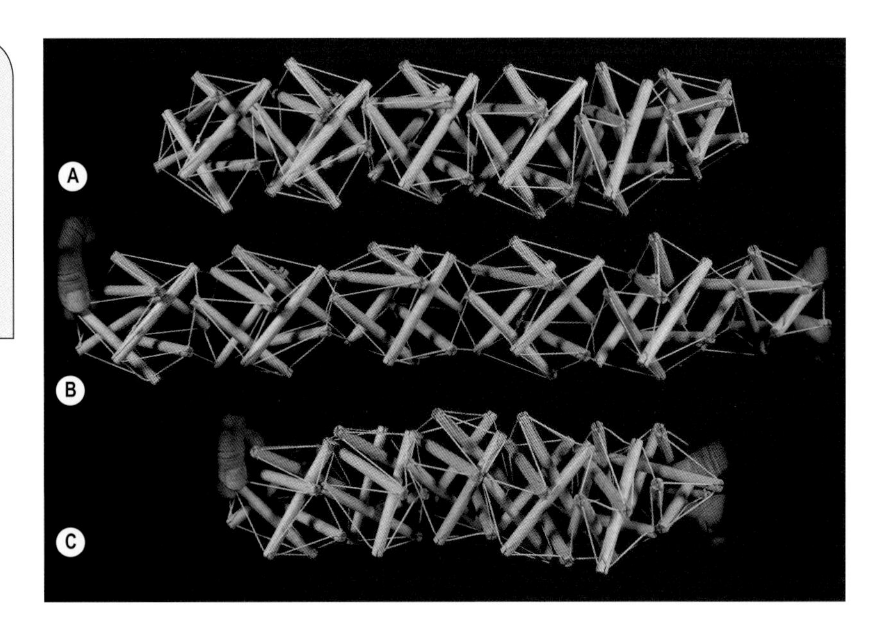

Figure 3.6
A A modular chain of tensegrity icosahedra can be stretched to form a cable B, or compressed to form a strut within an even larger hierarchical model C, © Rory James

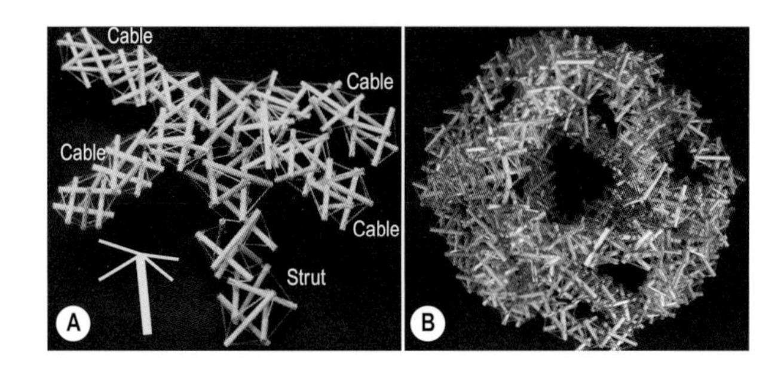

Figure 3.7
A A node (bottom left) with the strut and four cables modelled as a series of tensegrity icosahedra within a self-similar structural hierarchy; B a complete '6-strut' tensegrity-icosahedron with each cable and strut made from chains of smaller T-icosahedra © Rory James

Structural hierarchies

Structural hierarchies are an inherent capability of tensegrity configurations as well as being virtually ubiquitous in biology (Fuller, 1975, 740.00). They reduce weight, provide an energy-efficient means for packing components (Lakes, 1993), dissipate potentially damaging stresses (Gao et al., 2003; Gupta et al., 2006)

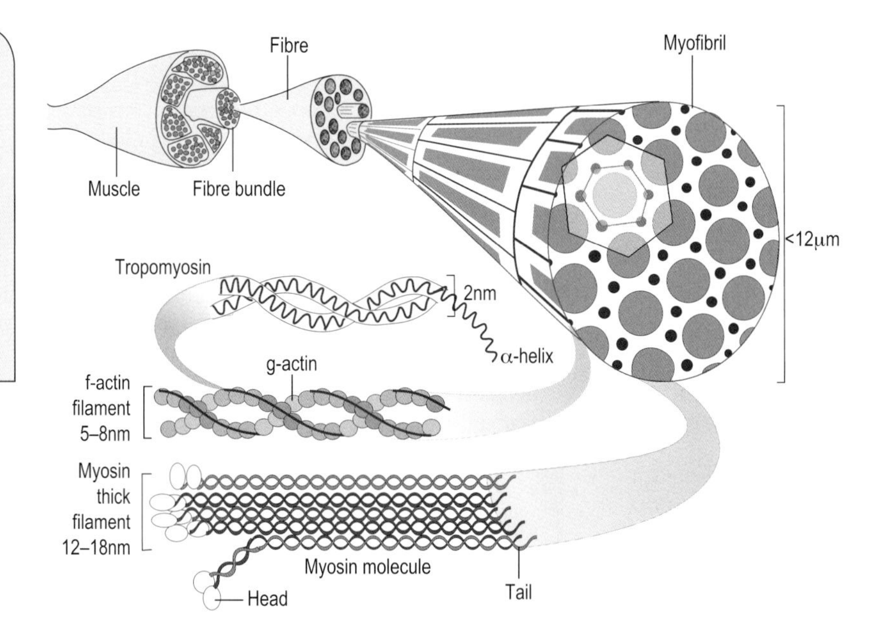

Figure 3.8
A structural hierarchy within muscle; note the helixes within helixes and the hexagonal close-packing arrangement of actin and myosin helixes within the myofibril. Reproduced with modifications from © Scarr, 2010, Elsevier

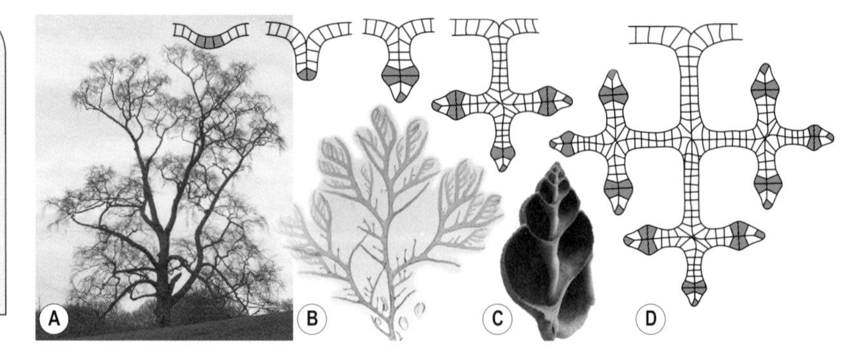

Figure 3.9
The self-similar fractal-like nature of: A a tree; B seaweed; C gastropod shell; D tissue formation through selective cellular growth (re-drawn from Huang, 1999).

and functionally connect every level from the simplest to the most complex, with the entire system acting as a unit (Figure 3.8). Simple patterns and shapes then repeat themselves at multiple size scales because they are the most energy-efficient configurations, and such fractal-like self-similarity underlies all growth processes in nature (Weibel, 1991; Jelinek et al., 2006) (Figure 3.9), with the branching of plants, bronchial (Weibel, 2009) and arterial 'trees' (Zamir, 2001) being well-recognised examples.

A pattern for all the others

At the atomic level, interactions between the forces of attraction and repulsion cause atoms to spontaneously assemble into the most stable and energy-efficient configurations, such as rigid crystals; and the Platonic shapes and isotropic vector matrix then showed how they can increase in size and retain the same basic shape (Figures 1.7, page 7; 2.7, page 16; 2.10, page 17 and 2.14, page 20).

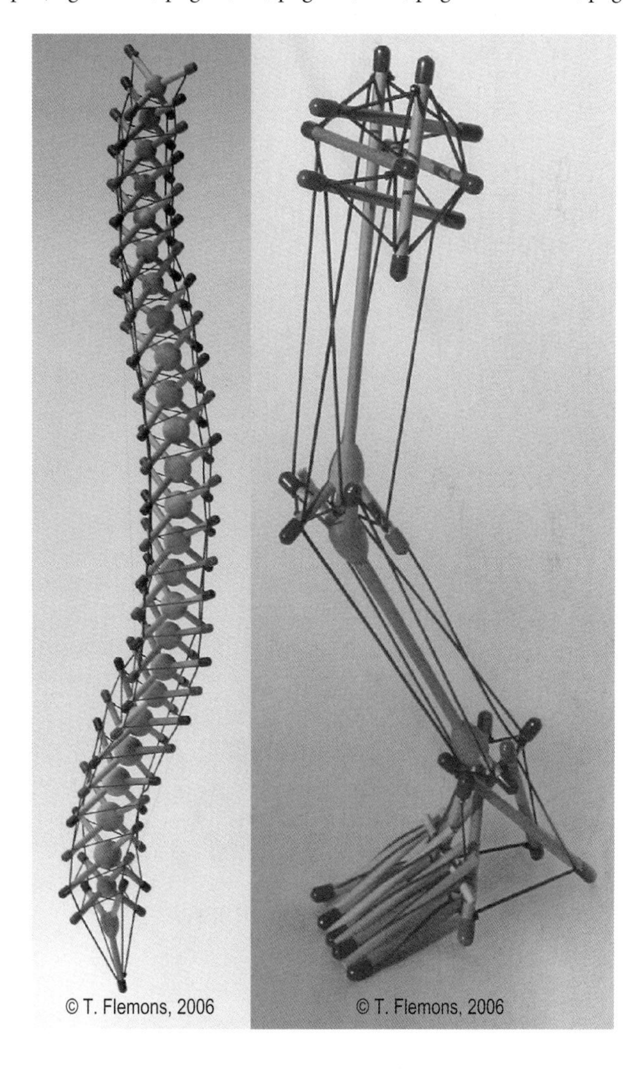

Figure 3.10
Tensegrity models as representations of the spine and leg. Reproduced courtesy of © T. Flemons, 2006 (Flemons, 2007, 2012)

Molecules are also influenced by the same rules of assembly and are tensegrity structures in their own right (Zanotti & Guerra, 2003; Edwards et al., 2012); and their flexibility allows them to join together to form even more complicated structures at higher size scales.

Of course, the same physical laws that govern the self-assembly of simple geometric shapes (Denton et al., 2003) are also going to influence the formation of more complex biological systems (Fuller, 1975, 220.04; Levin, 1995; Ingber, 1998), but using conventional methods to follow the interplay between all the forces within them is near impossible, which is why these stick-and-string models are so important. They enable the similar non-linear dynamics of tensegrity structures and biological tissues to be compared across multiple size scales, *no matter how complicated they become*!

The next few chapters provide some examples of how simple tensegrity models are changing our understanding of complex biology (Figure 3.10); and Levin emphasised that the reason they make so much difference is that tensegrities are not constrained by the man-made rules of classical mechanics, and neither is biology (Levin, 2006).

4 The problem with mechanics

Biological structures are chaotic, nonlinear, complex and unpredictable by their very nature... (Levin, 2006, p. 79)

Tensegrity is different to other man-made structures because of its unusual combination of features. Although Snelson's sculptural developments and Fuller's writings have been the inspiration that has carried it forwards, tensegrity has largely remained in the realm of the interested specialist and it has thus taken some time for it to become widely recognised.

As an architect, Buckminster Fuller persevered with his system of synergetic geometry and concentrated on the geodesic dome as a strong, cheap and lightweight alternative to masonry constructions but he also considered them as tensegrities, which has caused a great deal of controversy (Fuller, 1975, 703.03). The shape of a geodesic dome is based on the icosahedron and remains stable because of its configuration of triangulated struts (Figures 1.7 and 1.8, pages 7 and 8), which 'lock-in' the forces of tension and compression and distribute them throughout the structure. Although this balance of forces is part of what characterises tensegrity, and the atoms within the geodesic dome structure can certainly be considered as such (Connelly & Back, 1998), it is clear that the overall configuration is quite different at the macro scale and the debate on whether or not it is a tensegrity structure will probably remain for some time.

Structural and mechanical engineers are now taking more of an interest in stick-and-string tensegrity because of its remarkable characteristics (Motro, 2003; Jauregui, 2010), and are producing bridges and robotic structures for use in the exploration of space, etc, but these are often hybrids made from 'partial' tensegrities joined to standard construction designs. *True* tensegrities are made from multiple components linked in such a way that each has an influence on the position of *all* the others during movement, but the mechanisms needed to control them are still being developed and require solutions to some pretty complex calculations (Juan & Tur, 2008; Skelton & de Oliveira, 2009). Essentially, the simple rules of classical mechanics are not easily transferable to tensegrity.

The laws of classical mechanics

Classical mechanics is governed by a system of laws that determine what happens when an object is acted on by physical forces, and these laws have largely developed out of observations made since the sixteenth century by Galileo Galilei (1564–1642) and Isaac Newton (1642–1727), amongst others. However, while classical mechanics forms the basis for most man-made engineering projects, it has many shortcomings when applied to biological tensegrity.

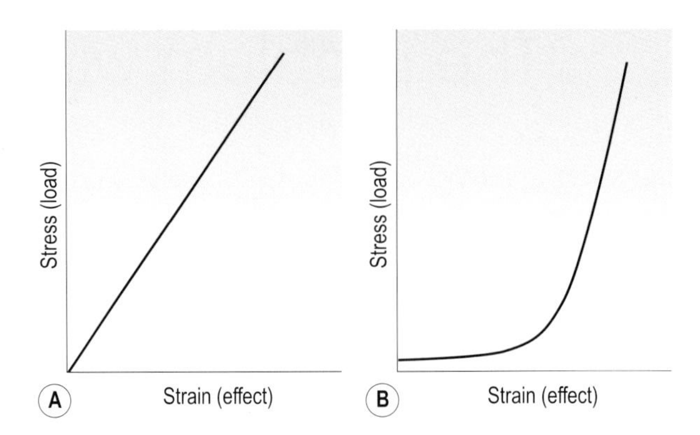

Figure 4.1
Typical stress-strain curves for materials that follow: A classical mechanics; B the non-linear dynamics of tensegrity and biological structures

Stress and strain

Sometime during the seventeenth century, Robert Hooke (1635–1703) was investigating what happens when a particular material is stretched or compressed. He formulated a law that effectively states that the amount that the material changes in length (the strain) is directly proportional to the force applied (the stress), i.e. there is a straight line, or linear, relationship between stress and strain (Figure 4.1a); which might seem obvious at first glance but this is not always the case. Tensegrity models and biological tissues differ from this in that their stress/strain curves are both *non-linear* (Figure 4.1b), and because of this similarity, it is worth considering them alongside each other.

Scaling up in size

Galilei had already presented his 'square-cube law' as a means of describing what happens when an object is scaled up in size, for example, where doubling the height of a building causes the surface area that it stands on to be squared and its volume (and mass) to be cubed (Galilei, 1638) (Figure 4.2). This law explains why the base of a skyscraper that is double the height of the one next to it must be strong enough to withstand an *eight-fold* increase in weight ($2^3 = 2 \times 2 \times 2 = 8$); but this doesn't necessarily follow in biology.

Large animals such as dinosaurs would have collapsed under their own weight if they followed this rule, and this anomaly has led to an apparent paradox.

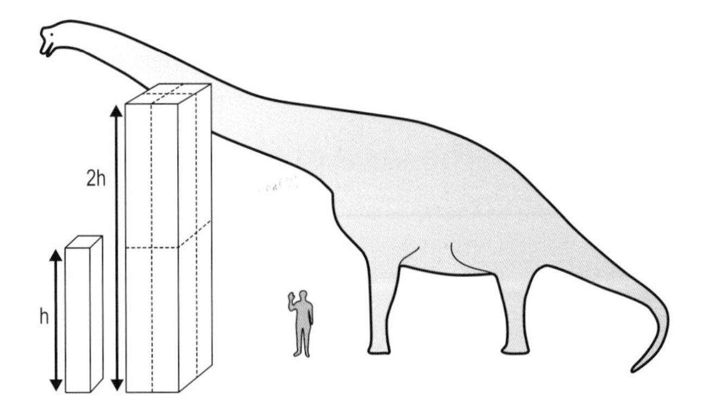

Figure 4.2
A proportional increase in size that doubles the height of an object causes the surface area to be squared and the volume (mass) to be cubed; but the leg bones of a large dinosaur such as *Giraffatitan* would have been unable to support its enormous weight (< 60 metric tons) if it followed Galilei's law.

Scaling up in size from smaller animals causes the calculated volume and mass of dinosaurs to increase at a far higher rate than the cross-sectional area of their leg bones; but because the compressive strength of bone is roughly the same in all animals, their legs would simply not have been strong enough to support this huge increase in weight (Levin, 2006, p. 69). This means that either dinosaurs could not have survived for long or that the square-cube law doesn't necessarily apply to these large animals and, as the fossil evidence suggests that dinosaurs were a very successful and dominant group of animals over millions of years, the latter is probably more likely to be the case. Even though they do have bigger bones, it may be that body mass is not the primary driver of bone shape (Fabre et al., 2013).

The consequences

If the behaviour of living organisms were to follow the laws of classical mechanics, then lifting a heavy weight would cause muscles to tear, discs to rupture, vertebrae to be crushed and blood vessels to burst; fly fishing would become a dangerous sport and simple everyday activities would be fraught with all sorts of problems (Levin, 2002). In addition, Levin noted that the uterus and bladder would burst when they contracted, the heart would be unable to maintain the systolic pressure and the bones of humans and larger animals would be likely to fracture during movement.

A glimmer of hope

Of course, healthy tissues rarely collapse and this is partly because they are *not* constrained by these laws. Most man-made materials become weaker when they are stressed, and there comes a point when they start to deform and break apart if the force is increased further, which means that they are often designed to function safely at relatively low strain levels (sometimes as little as 10%) (Gordon, 1978, p. 149) (Figure 4.1a). Tensegrities and biological tissues are different in that they can resist much higher strains for their size and weight (sometimes up to 100%) – as the initial part of the curve in Figure 4.1b indicates – and they actually become stiffer and stronger as the force increases.

Bones and tendons can thus store much larger amounts of energy and return it like a spring (Biewener, 1998). Tissues under high strains can remain flexible, tough and completely stable, and structures that would burst or collapse if restricted to following the laws of classical mechanics remain completely stable (Gracovetsky, 1992; Levin, 2006, p. 76). In addition, the intrinsic tension (pre-stress) contained in both tensegrity models and biological structures means that the initial part of the stress/strain curve never reaches zero and the structure is always 'primed for action' (Gordon, 1978, p. 156).

This relationship between stress and strain (cause and effect) is the mainstay of classical mechanics, but it can take different forms. For example, Siméon Poisson (1781–1840) noted a particular stress/strain relationship (Poisson's ratio) where most man-made materials get thinner when stretched and bulge in the middle when compressed, in a ratio that is typically between 0.2 and 0.5; i.e. the amount of strain is between 20% and 50% of the stress but perpendicular to it.

However, tensegrity models and biological tissues frequently do the opposite, with the structure getting *thicker* when stretched and *thinner* when compressed (a 'negative' ratio), with values that typically approach 1.0 (Gordon, 1978, p. 161). The omni-directional expansion and contraction of the T-icosahedron (jitterbug) is an important example of this (Fuller 1975, 724.31); Levin, 2002) (Figures 2.18, page 22 and 3.4, page 28).

The important bits are missing!

Part of the problem with describing biology through classical mechanics is that these 'laws' (of which there are many more) were discovered through experiments on inanimate objects, and it has generally been assumed that the 'odd' behaviour of living organisms was something that would eventually 'fall into place' or that could be conveniently ignored. However, the results of cadaver experiments are frequently based on assumptions that cannot be substantiated (Van der Wal, 2009) and the addition of 'just a bit more data' misses the point because living tissues behave differently in any case. In addition, an analysis that only considers the relatively straight high-strain region of the non-linear curve, and disregards the initial part, will be misleading (Figure 4.1b) because it is *not* representative of tissues that are functioning at their normal mechanical and energy-efficient optimum. Although classical mechanics is essential to engineering and has played a significant part in the development of medical devices, etc. it is important because of its ability to describe the relationship between an object and the forces acting on it in *3-dimensional Euclidean space;* but biology is not constrained by Euclidean geometry (Lighthill, 1986; Levin, 2002; McFadden, 2013).

A different sort of geometry

Euclid was a Greek mathematician who lived sometime around 300 BCE and is noted for his development of a logical system of planar geometry, which was so intuitively correct that any theorem derived from it was deemed to be an absolute representation of an unchanging reality. In fact, Euclidean geometry remained essentially unchallenged until the late nineteenth and early twentieth centuries, when quantum mechanics and Einstein's theories of relativity, etc. revealed that there are other ways of looking at geometry and space (see Chapters 9 and 10), but it is still taught in schools today and remains part of our everyday experience (Fuller, 1975, 203.04). Classical mechanics thus developed within this spatial system because it was the only one that could be recognised at the time, and its application to biology was inevitable, but it has led biomechanics down a blind alley from which it is only just starting to recover (Levin, 2002).

Biomechanics

Orthodox biomechanics is based on a system of pillars, beams and cantilevers (like skyscrapers and bridges), where the bones are stacked on top of one another like a pile of bricks (Figure 1.5, page 6) and the muscles and connective tissues control them like the rigging in a sailing ship – the 'columns and arches' support the weight, and the soft tissues move them about in a local piecemeal-like way.

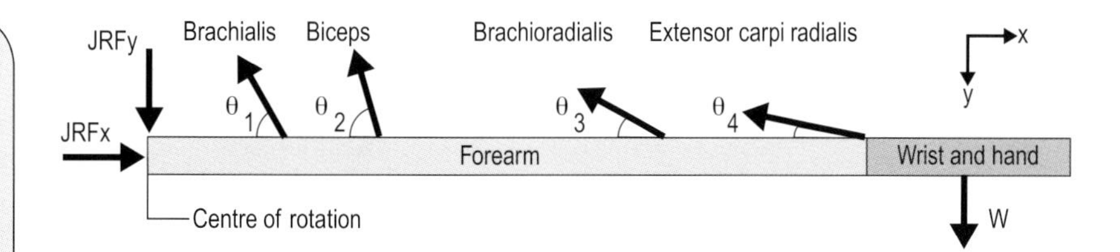

Figure 4.3
A typical free-body diagram of the lower arm showing the forces resulting from the particular orientations of muscle insertions, the two components of the joint reaction forces at the elbow (JRFx and JRFy) and supported weight (W). This kind of analysis neglects the mass of the forearm, the effect of force changes during movement, the influence of polyarticular muscles and connective tissues, and erroneously assumes that the JRF is effective about a fixed centre of rotation, etc. (re-drawn from Ethier and Simmons, 2007).

Motion is then analysed through a system of joints and levers, where the bones compress each other and each joint is considered in isolation from the others and the connective tissues that surround and link them together (Standring, 2005, p. 134) (Figure 4.3); but such analysis is a gross simplification of joint function, always incomplete and in some cases, absurd (Gracovetsky, 1992; Levin, 2002).

The broken lever

Simple levers are machines that consist of a pair of beams or 'arms' that move in relation to each other through the application of some force. Their advantage is in the amplifier effect where a small load, positioned some distance away from the fulcrum, can be balanced by a much larger force that is closer to it (Figure 4.3); and the actual amount needed to balance or move a particular load is then just a simple calculation. Figure 4.4, for example, shows the amount of force needed to balance a two kilogram fish at the end of a three-metre long fishing rod; but it soon becomes clear that the muscles and other soft tissues in the back would be unable to support this (Hansen et al., 2006). Such 'free-body' analysis assumes that the spine is a single column, when in fact it is a hugely complex arrangement of bones, muscles and connective tissues that could be severely damaged if they were part of a lever system (Levin, 2002).

Levers generate bending moments and stress concentrations, and it is likely that such tissues would collapse through material fatigue at an early stage in development, and that really would be the end of them (Levin, 2006, p. 72) (Figure 4.5). While skyscrapers and bridges are relatively rigid and must be heavily reinforced against the effects of these stresses, living organisms are light and flexible, and can function the same in virtually any position, which means that they probably use a different system of mechanics altogether.

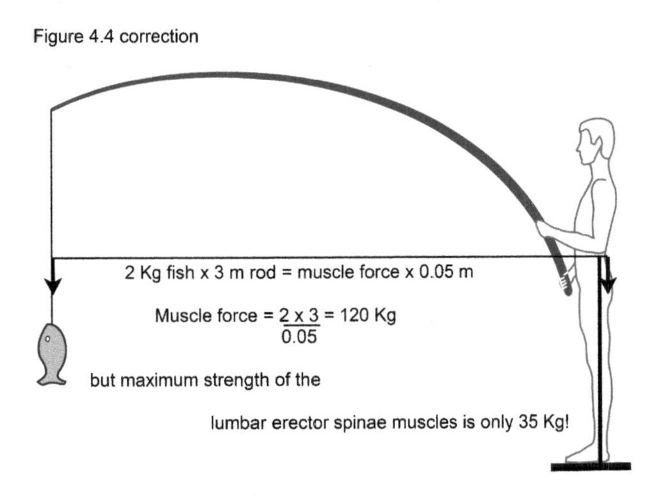

Figure 4.4
A free-body diagram of a lever system showing that a 2 kg fish at the end of a 3 m rod would require a balancing force that would exceed the strength of the back muscles (re-drawn from Levin, 1995).

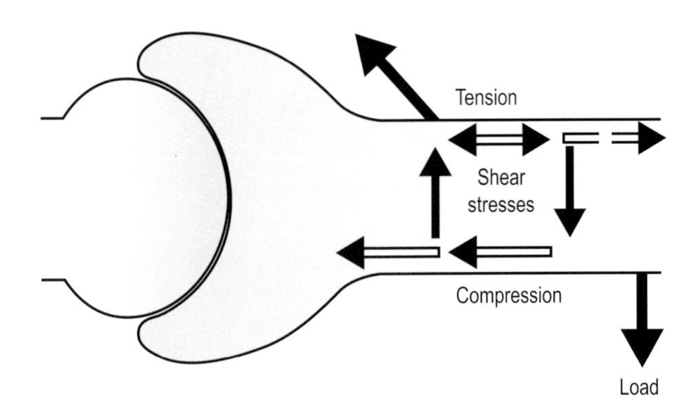

Figure 4.5
Levers generate bending moments and potentially damaging shear stresses (vertical arrows) within the structure itself, as loading creates tension and compressional forces (open arrows) that cause one side of the bone to be stretched and the other side compressed.

A changing paradigm

As an orthopaedic surgeon, Stephen Levin found that normal bones do *not* compress each other across their joint surfaces and appear to 'float' within the soft tissues (Levin, 1981); and it has also been known for some time that bones do not revolve around fixed axes but follow helical trajectories that are constantly changing position during movement (Chapters 6 and 7) (Standring, 2005, p. 110). Classical mechanics and isolated lever systems are simply not good enough to explain the complexity of biological motion and there is now an alternative that provides a more thorough description. There is nothing wrong with classical mechanics but its value in describing biology is now largely outdated (Levin, 2002)!

Bio-tensegrity

Humans function much the same in *any* position, irrespective of the position of gravity and, from an energetic perspective, a tensegrity system that integrates every part of their anatomy through a pre-tensioned hierarchy would be much more attractive than having to rely on a stress-ridden system of isolated beams, pillars and levers (Turvey, 2007).

Biotensegrity systems consist of just two components, tensioned 'cables' and compressed 'struts', with the stick-and-string models demonstrating every aspect of their dynamics; they are strong, light in weight and flexible, and can change shape with the minimum amount of effort. This change in position of one part relative to another is part of what characterises movement, and must happen in a controlled way if it is to be purposeful, but it is well known that the nervous system is incapable of regulating every aspect of this (Stecco, 2004, p. 31). Tensegrity models show that the structure *itself* is capable of controlling complex movements, and this is because their mechanics are based on the geometric relationships between *all* the different parts. Kinematics, as the physical analysis of the geometry of motion, then explains how the forces of tension and compression can be transmitted in a definable way (Muller, 1996).

The kinematic chain

Although kinematic analysis is considered a branch of classical mechanics, it still has its uses in describing biology, where multiple joints are linked into a

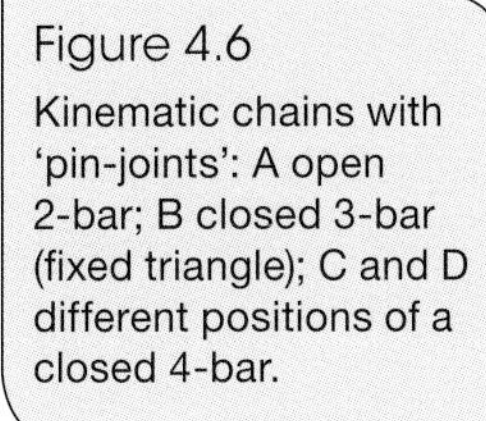

Figure 4.6
Kinematic chains with 'pin-joints': A open 2-bar; B closed 3-bar (fixed triangle); C and D different positions of a closed 4-bar.

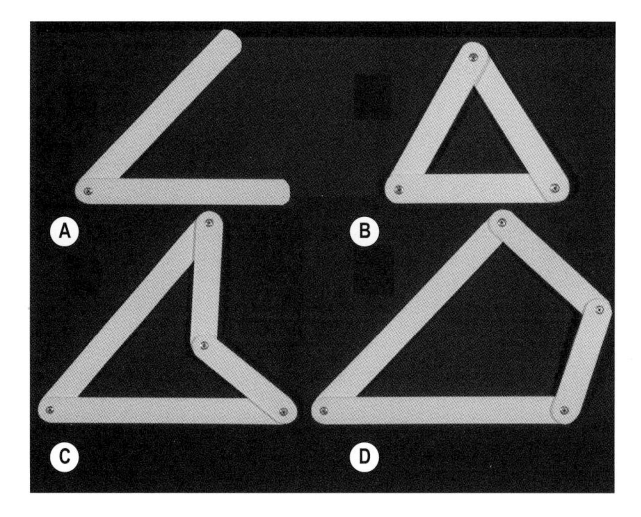

unified movement system that is partly controlled by the structure itself. The simple 2-bar lever is also related to this (Figure 4.6a), but as we have already seen, it has many inadequacies from a biological perspective because it is an isolated and *open* system.

Closed-chain kinematics

Closed kinematic chains are different in that they couple multiple joints into a *continuous* mechanical loop that allows motion, with the movement of each 'joint' causing controlled changes in position, velocity and kinetic energy at *all* the others, and with a definable relationship that depends entirely on the geometry (Figure 4.6c and d) (Alfaro et al., 2004). A typical example of these can be seen in the traction mechanism of a steam locomotive (Figure 4.7).

Another application can be observed in the *Strandbeest* produced by Theo Jansen, a constructivist artist in the Netherlands who connects many kinematic chain mechanisms into wind powered 'creatures' that walk along the beach (Figure 4.8).

The simplest of these closed-chains (Figure 4.6b) has three bars and three joints and forms a rigid triangle – most important to biological mechanics because of its inherent stability and contributor to the tetrahedron, octahedron and icosahedron, the basis of all natural shapes (Chapter 2). To be fair to the lever, we could also think of one of the bars as a compliant muscle that moves the two arms of Figure 4.6a, but this would still remain an isolated system on its own.

Of course, biological 'joints' are not pin-joints (Figure 4.6) but part of enormously complex and interconnected hierarchies (Figure 3.7, page 30), with each 'bar' made from numerous *smaller* 'bars' that have different mechanical properties and further influence how they respond to the forces of tension and compression. Each bar could be a passive structure such as a bone or tendon, or an active motion generator such as a muscle; and a 'joint' in this hierarchical context also refers to the attachment of one particular tissue to another such as fascia to bone, or muscle fibres to an aponeurosis, etc. rather than just a 'knee' or 'elbow'.

Figure 4.7
The traction mechanism of a steam locomotive where a single drive on either side of the engine is coupled to the wheels through several closed and interlinked kinematic chains.

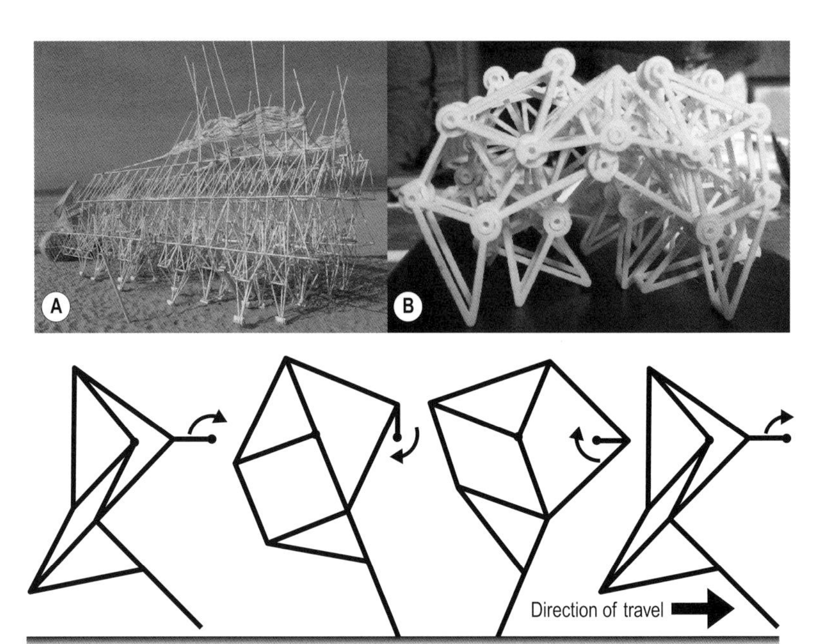

Figure 4.8
A Theo Jansen *Strandbeest* 'creature', reproduced courtesy of © Theo Jansen; B a similar model showing the many interlinked kinematic chains (courtesy of Theo Jansen); C the basic 3 and 4-bar kinematic chains that control the changes in leg shape and movement.

Tensegrity kinematics

Levin pointed out that tensegrity cables and struts *also* form the bars of closed-chain kinematic systems, and that they fill space in every direction, with the '3-bars' providing the relative stability that allows the '4-bars' to change shape (Swartz & Hayes, 2007; Levin, 2014) (Figure 4.9). The T-icosahedron 'jitterbug' is a good demonstration of this (Fuller, 1975, 724.31) (Figure 3.4, page 28).

Pushing opposite 'tension-triangles' together causes the equatorial nodes to twist and the overall structure to contract (within the constraints of the materials used), and releasing the pressure then allows it to return to its normal state of equilibrium. The jitterbug actually represents transformations in energy that oscillate from one state to another, and because tensegrity structures are complete energy systems in their own right, contributes to the T-icosahedron as

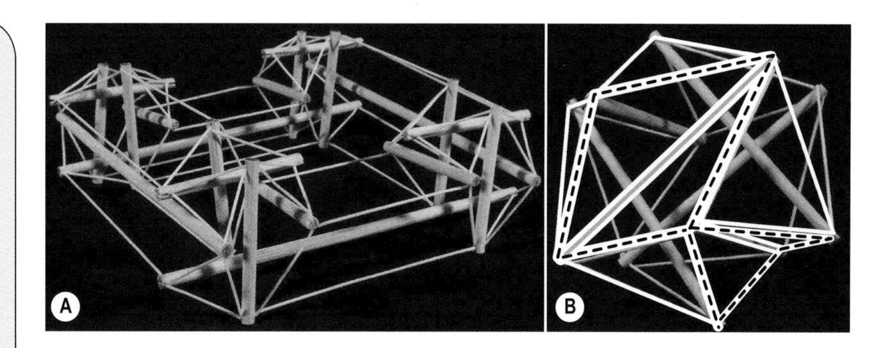

Figure 4.9
A An oblique view of a closed 4-bar kinematic chain (Figure 4.6 C and D) modelled as a tensegrity hierarchy made from four inter-linked T-icosahedral 'joints'; B a T-icosahedron showing some of the stable 3-bar (white lines) and flexible 4-bar (dotted) kinematic chains that guide changes in shape. © Rory James

the archetypal model to represent biological structure and motion (Fuller, 1975, 724.32; Edmondson, 2007, p. 179) (Figure 2.18, page xx).

The problem solved!

Biology is not constrained by the laws of classical mechanics and, if there is to be a genuine understanding of its functions, an organism must always be considered in its entirety with each 'part' related to the whole. Fuller recognised that an appreciation of basic physics is essential to explaining the complexities of biological structure, and that tensegrity linked them both together (Fuller, 1975, 707.03). Levin then explained biotensegrity from first principles and a hierarchical perspective (Levin, 1981, 2006); and Donald Ingber (b. 1956), a cell biologist investigating the role of the cellular cytoskeleton in angiogenesis, considered the tensegrity concept through experiment and confirmed its importance to cell function and the hierarchical body (Ingber et al., 1981; Ingber, 2008). We can now look at some examples of how biotensegrity is improving our understanding of human anatomy – a dynamic architecture that unifies the whole organism.

5
The autonomous cell

The beauty of life is... that of geometry with spatial constraints as the only unifying principle. Donald E. Ingber (Edmondson, 2007, p. 286)

As an undergraduate in the 1970's, Donald Ingber was attending a sculpture class in three-dimensional design and, impressed with the tensegrity structure that was displayed in front of him, he recognised similarities with the behaviour of cells cultured in a petri dish. The tensegrity model flattened when pressed onto the table and rounded up when released, and the flattened cells on the bottom of the dish did the same thing when an enzyme was added to degrade their sticky adhesions. Ingber then realised that cells might also be tensegrity structures (Ingber, 1993).

The cytoskeleton

Originally thought of as a membranous bag of gel, the internal structure of cells is now known to consist of a framework of molecular microfilaments (MF), microtubules (MT) and intermediate filaments called the cytoskeleton. It is like a tensegrity scaffold that mechanically links different parts of the cell and influences its shape and function (Figure 5.1) (Ingber, 2003a). Ingber reproduced the behaviour of cultured cells by building a simple tensegrity model of his own but with the addition of a smaller nuclear one inside that linked to the outer

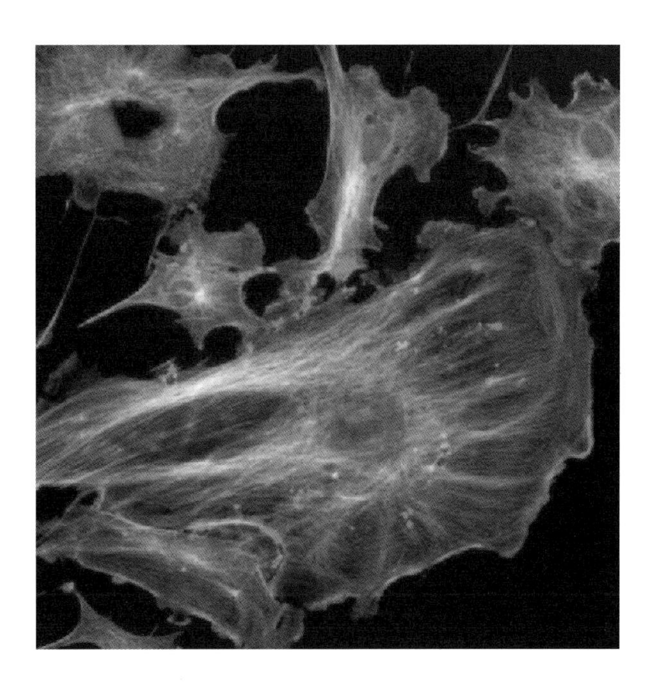

Figure 5.1
Endothelial cells from bovine pulmonary artery showing the cytoskeleton with its numerous microtubules (light) extending outwards from the middle of the cell, and finer actin microfilaments (public domain, reproduced from Wikipedia).

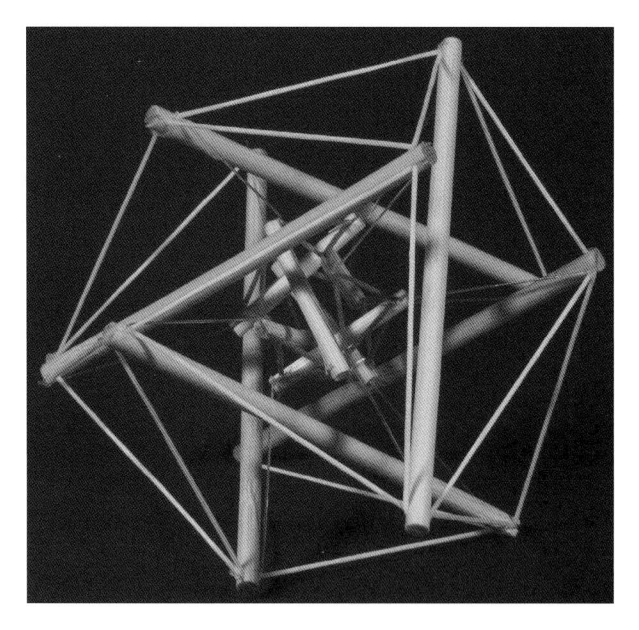

Figure 5.2
A tensegrity-icosahedron (T6-sphere) connected to a smaller one inside shows how changes in shape will affect them both. © Rory James

model, i.e. a structural hierarchy (Figure 5.2). By pressing down or pulling on opposite sides, the inner 'nucleus' was then caused to change shape and follow the outer part in a similar way. This behaviour was also demonstrated in living cells and it was later confirmed that the cytoskeleton does indeed connect the outer cell cortex with the nucleus and chromosomes within it and thus acts mechanically like the tensegrity model (Wang et al., 2009).

Regulating the cell

The different components of the cytoskeleton are connected so that tensioned microfilaments are balanced by microtubules under compression (Brangwynne et al., 2006), with tensioned intermediate filaments linking them all together from the cell membrane to the nucleus. They are all part of a dynamic structure that is constantly assembling and disassembling its molecular components in response to mechanical forces. Tension is generated through the action of actomyosin motors and polymerization of microtubules (Tolié-Nørrelykke, 2008), and any change in force at one part of the structure causes the *entire* cytoskeleton to alter cell shape – also a characteristic of tensegrity (Stamenovic & Ingber, 2009).

The cytoskeleton is further interesting because it is a multi-functional structure that influences the shape of the cell and position of organelles, activates multiple intra-cellular signalling cascades and has links to other cells and tissues (Mammoto et al., 2010). It also connects to the surrounding extracellular matrix (ECM) and other cells through adhesion molecules that cross the cell membrane, such as integrins and cadherins, respectively (Figure 5.3). These transmembrane proteins then create a mechanical coupling that transfers tension generated within the cytoskeleton to the ECM and adjacent cells and, because an intrinsic state of tension exists between them (pre-stress), so a change in ECM tension also causes a realignment of structures within the cytoskeleton and alters cell shape and function (Wang et al., 2009).

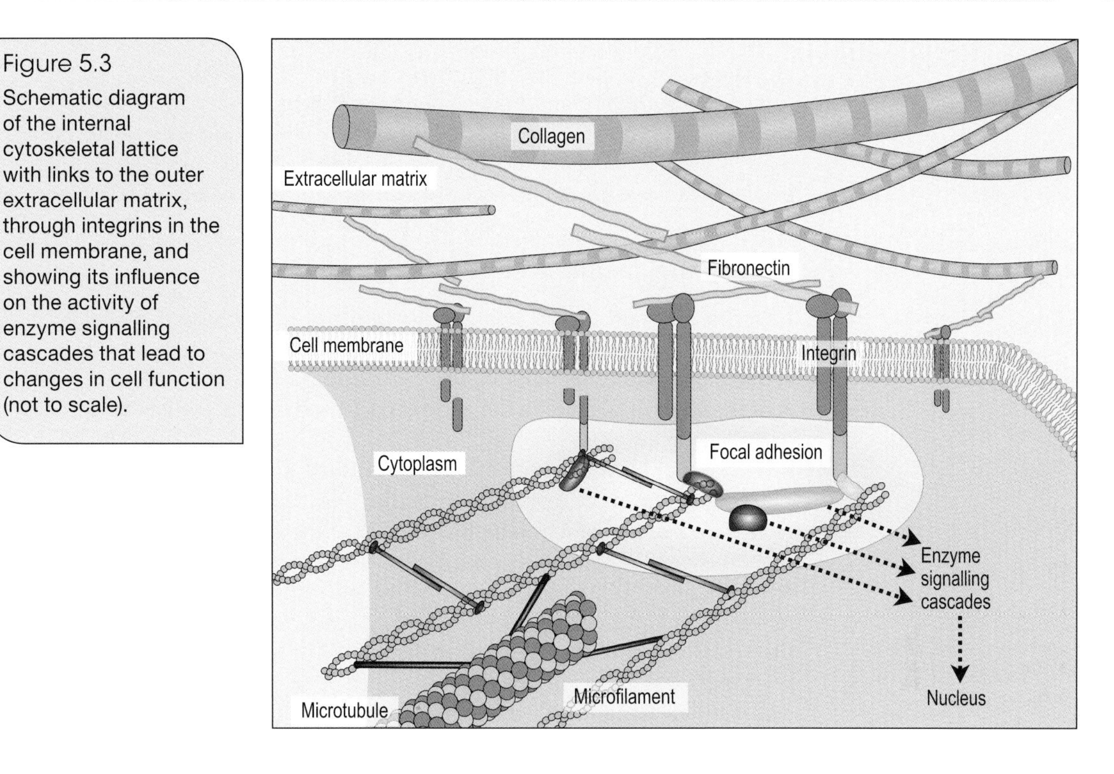

Figure 5.3
Schematic diagram of the internal cytoskeletal lattice with links to the outer extracellular matrix, through integrins in the cell membrane, and showing its influence on the activity of enzyme signalling cascades that lead to changes in cell function (not to scale).

Shaping the balance

Experiments that allowed individual cells to assume certain shapes by culturing them on specially prepared adhesive islands of circles, squares and triangles showed that those able to distort or spread had the highest rates of growth, rounded cells became apoptotic and died and those intermediate in shape became quiescent and differentiated (Ingber, 2008). Cells also tended to extend new motile processes, such as lamellipodia and filopodia, on sharp corners rather than blunt ones (Figure 5.4) (Parker et al., 2002). These findings have all contributed to our understanding of how cell shape, and links between the cytoskeleton, nucleus, metabolic substrates and DNA are able to alter cell function through a process known as *mechanotransduction*, where mechanical signals are converted into chemical ones (Wang et al., 2009).

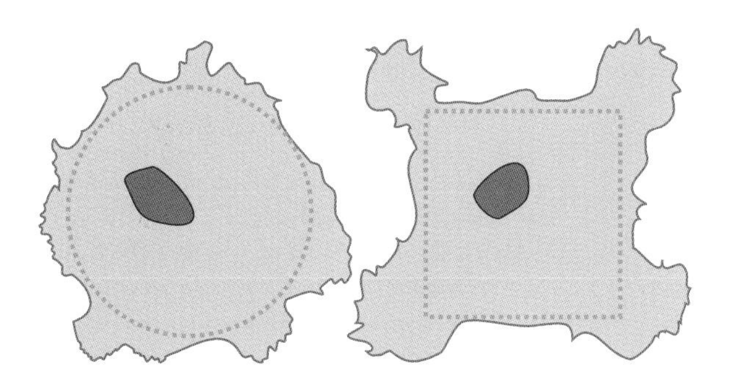

Figure 5.4
Endothelial cells preferentially extend cell processes outwards from corners, when grown on adhesive islands with different shapes (dotted).

Many enzymes and substrates are attached to the cytoskeletal lattice and mediate critical metabolic functions such as glycolysis, messenger RNA transcription and protein synthesis, and DNA transcription and replication are also carried out on nuclear scaffolds that are continuous with the rest of the cytoskeleton. Global changes in tension and cell shape then alter cellular biochemistry and ultimately lead to a switching between different functional states such as growth, differentiation and apoptosis (Wang et al., 2009).

Linking the 'inside' with the 'outside'

Integrins are really transmembranous strain gauges that respond to changes in tension on both sides of the membrane (De Santis et al., 2011), and their activation promotes the binding of other proteins that connect them to contractile actin bundles in the outer cortical part of the cytoskeleton and contribute to a specialised complex called a *focal adhesion* (Figure 5.3). The attachment of ECM molecules (e.g. fibronectin) to particular integrins then causes a change in actin tension within the cytoskeleton, and stimulates the recruitment of more focal adhesion proteins to the same area that together share the load (Ladoux & Nicolas, 2012). These changes in tension can then feed back to the ECM and cause unfolding and fibrillogenesis of the fibronectin molecule and ultimately the assembly of collagen (Kadler et al., 2008). Because the geometric spacing of fibronectin nanofibrils on the outside of the membrane is proportional to the spacing of cross-linked actin bundles within the cytoskeleton (Pompe et al., 2005), the cell is able to maintain tight regulatory control over collagen morphogenesis and maintenance of the ECM during development (Martin et al., 2000; Mao & Schwarzbauer, 2005).

The development of tissues

During embryogenesis, the reciprocal exchange of tension between the cytoskeleton and ECM coupled with the influence of chemical mitogens, can cause changes in cell activity and the production of enzymes that degrade the basal membrane (ECM) beneath the cell (DuFort et al., 2011) (Figure 5.5). As this leads to local variations in its thickness, the intrinsic tissue tension then causes the ECM to stretch and distort slightly (Huang & Ingber, 1999) leading to the development of more complex patterns such as budding, branching (e.g. alveoli) (Moore et al., 2005), and the formation of tubes (capillaries) (Huang & Ingber, 1989) (Figures 3.9a, b & d, page 30) or the transition from epithelium into motile mesenchymal cells that migrate away to a different region (Ingber, 2006). While the ability of cells to attach to an underlying substrate and exert traction forces has been known for quite some time (Harris, 1980), it is the biotensegrity model that has provided the most thorough explanation of their movement.

The movement of cells

Mesenchymal cells frequently move in response to chemical signals that diffuse through the ECM, and they do this by coordinating changes in the tensegrity cytoskeleton. Such cells attach themselves to the relatively rigid ECM through focal adhesion binding sites, a process that leads to actin filaments in the cytoskeleton stiffening and remodelling themselves into basal 'stress' fibres and

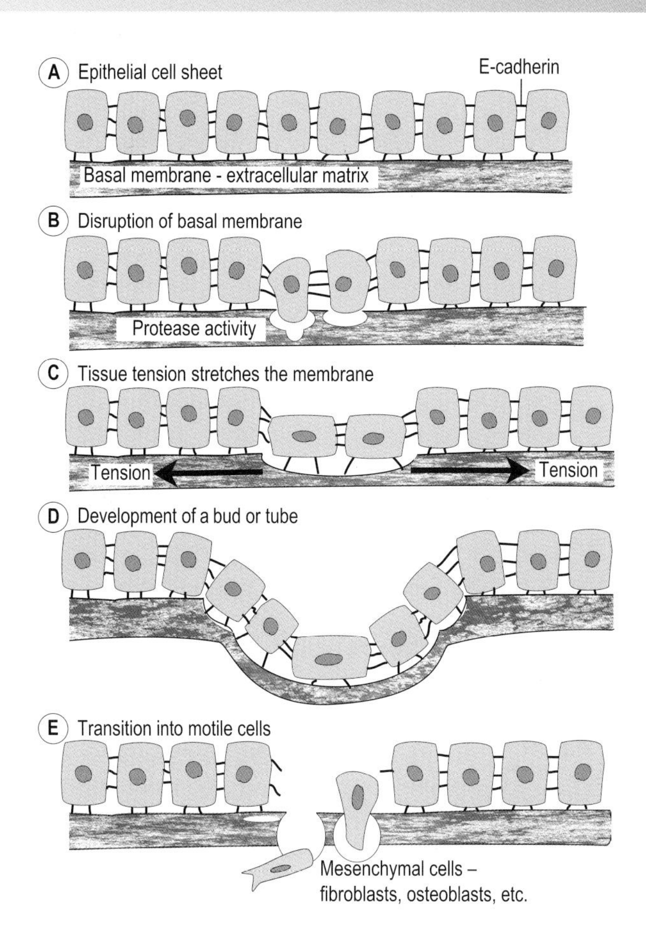

Figure 5.5
A sequence of changes that can occur in an epithelial sheet: A cells attached to each other and the tensioned basement membrane (BM); B changes in cytoskeletal activity lead to the production of enzymes that degrade the BM locally; C stretching of the thinner BM distorts the cells and further changes their function; D development of a bud, branch or tube; E epithelial-mesenchymal transition into motile cells (re-drawn with modifications from Huang & Ingber, 1999).

an overlying apical dome, with the resulting formation of a moveable filopodium that projects outwards (Figure 5.6b). Further changes in tension within the basal cytoskeleton then produce new adhesion sites near to the leading edge, which anchor the cell and lead to the formation of a second filopodium, as the process repeats and the cell pulls itself forward (Stamenovic & Ingber, 2009).

Throughout this process, tensioned actin filaments (cables) in one part of the cell are organising themselves into bundles of basal stress fibres and effectively change into tensegrity struts that contribute to the formation of filopodia. In other parts of the cell, different bundles are also disassembling and reverting to their previous state. The cycle continues as the cytoskeleton responds to local changes in tension, and the tensegrity configuration propagates the effects of these throughout the cell, but there can come a point when the mechano-chemical environment changes and the cell switches function once again.

The development of complex patterns

Living cells are dynamic systems, with an internal environment that is in a constant state of flux and a tensegrity cytoskeleton that is continuously trying to balance itself. The transfer of mechanical forces between the cytoskeleton

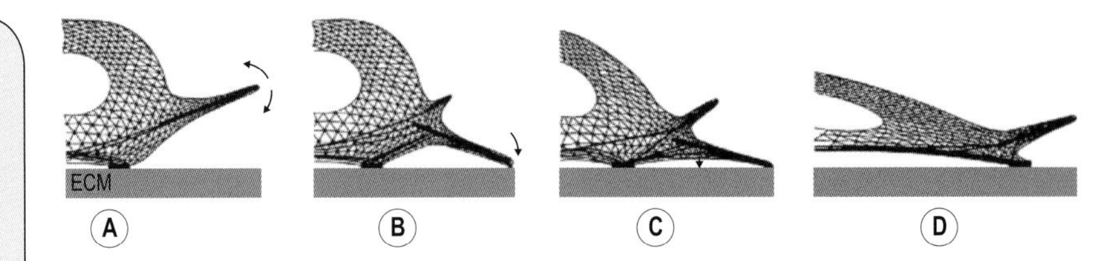

Fig. 5.6
A Attachment of a cell to the ECM causes actin fibres in the cytoskeleton to stiffen and remodel into basal 'stress' fibres, with the formation of an apical dome and moveable filopodium; B a new attachment to the substrate changes the balance within the cytoskeleton and leads to the development of a new filopodium; C the tensioned actin net is pulled forwards along the original stiffened filopodial core as the new filopodium develops and the basal stress fibres connect with those further back; D the filopodial core merges with the rear basal stress fibres and a new attachment to the ECM is formed as the cell pulls itself forwards. Reproduced from © Ingber et al, 1994, Elsevier.

and ECM can lead to an increase in cell growth and an expanding cell population, which then imposes stresses on surrounding tissues and influences their own development (Chen & Ingber, 1999; Rozario & DeSimone, 2010). Whenever two adjacent tissues grow at different rates, the rapidly growing one will compress the slower one and cause it to be stretched under tension (Henderson & Carter, 2002; Blechschmidt, 2004), with the resulting outcome dependant on the tissue compliance and cytoskeletal response of its own cells (Figure 5.5). Complex multi-cellular tissue patterns can then emerge some distance away, with their formation based on the same principles (Nelson et al., 2005) and continuity of the ECM with the fascial network could extend this throughout the developing embryo.

The cellular integrator

The role of the cytoskeleton as an intrinsically tensioned *tensegrity* network that influences the functional activities of cells is now becoming more widely accepted in scientific circles, and is largely due to developments in cellular biochemistry, microscopy and computerised modelling (with closed-chain kinematic analysis playing a part in this) (Chen et al., 2010), but the inspiration was provided by the simple stick-and-string models.

The transfer of forces through a network that coordinates the development and functional activities of cells (Vogel & Sheetz, 2006; Jamali et al., 2010), tissues and organs as parts of a globally tensioned unit (Chen & Ingber, 1998; Parker & Ingber, 2007; Mammoto & Ingber, 2010) would also be an efficient mechanism for controlling whole-body movement (Levin & Martin, 2012); but in order to examine this in more complex structures, we need to reintroduce some of the basic geometry, starting with the helix.

6 The twist in the tale

Nature may be a game of chance, but it plays with loaded dice.
Stephen M. Levin (2014)

The cytoskeleton is a most important example of biotensegrity in action because it clearly links structural mechanics with cell function, and as we have already seen, it is the geometry that underlies the mechanics. Nature always does things in the most efficient way, and it is no surprise that simple geometric shapes frequently appear in biology and repeat themselves at multiple size scales. One such shape is the helix.

The helix

At the most basic level, the tetrahedron is the shape that encloses the smallest amount of unit space within a given surface area. The tetrahelix that develops from it is efficient in the same way but it still cannot fill space completely (Figure 6.1) (Lord & Ranganathan, 2001a). Most (if not all) molecular helixes are geometrically related to the tetrahelix, and although it is the *simplest* of all helixes, it already contains an intrinsic hierarchy with its three sub-helixes of different pitch and chirality (Figure 6.2) (Sadoc & Rivier, 2000; Lord 2002); even the double-helix of DNA has been described as a tetrahelix with one of the long strands missing (Sadoc & Rivier, 2000; Fuller, 1975, 931.63). The helix is a common motif in protein structure and a general model for coiled winding in a wide variety of different organisms and structures at many hierarchical levels (Scarr, 2011).

The molecular helix

Biological helixes are usually thought of as being continuous, like metal springs, but of course they are nothing of the sort; they are made from many individual components connected together, just like tensegrity models (Figure 6.1). Molecules

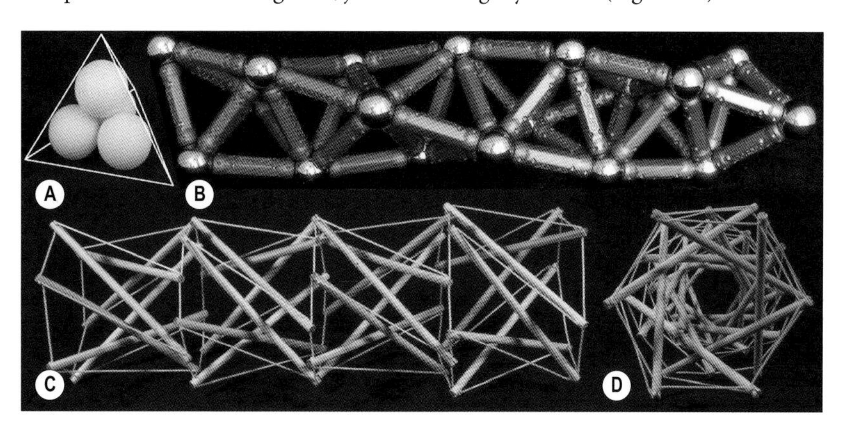

Figure 6.1
A model for biological helixes: A a tetrahedron encloses the smallest amount of unit space; B tetrahelix; C and D a T6-helix (side and end views). © Rory James

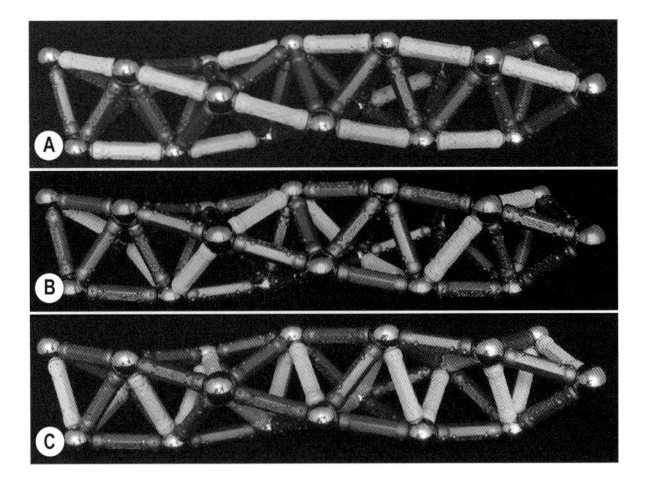

Figure 6.2
A tetrahelix (dark) intrinsically contains three sub-helixes of different chirality and pitch (light).

are thus tensegrity structures in their own right, because they automatically balance the forces of attraction (tension) and repulsion (compression) between their constituent atoms (Zanotti & Guerra, 2003; Edwards et al., 2012), and spontaneously assemble themselves through the inter-related principles of geodesic geometry, close-packing and minimal-energy (Fuller, 1975, 740.21).

Many helical molecules are made from globular proteins that contain even smaller helixes, such as g-actin and tubulin in the cytoskeleton (Figure 6.3).

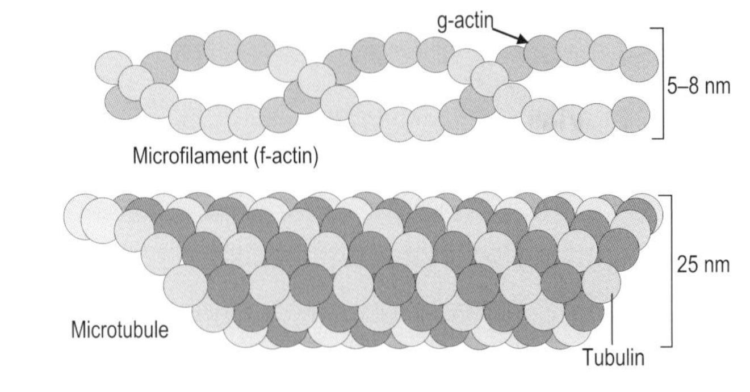

Figure 6.3
Helical molecules in the cellular cytoskeleton (re-drawn from Scarr 2010).

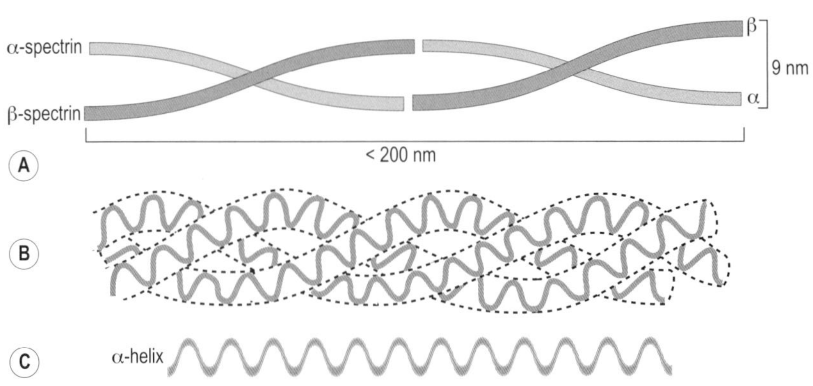

Figure 6.4
Schematic diagram of the spectrin hierarchy: A a spectrin filament basically consists of a series of tetramers made from α and β-spectrin dimers; B each dimer is formed from a series of double and triple coiled-coils of C spectrin α-helixes (re-drawn from Scarr, 2010).

Similar helixes can also wind around each other to form coiled-coils (Parry et al., 2008), such as filamentous spectrin (Figure 6.4), or further combine into more complex hierarchical structures with specialised functions, such as vimentin (Qin et al., 2009) and collagen (Fratzl, 2008).

Complex hierarchies

Spectrin

Spectrin filaments fixed to ankyrin in the lipid membrane are coupled to underlying bundles of actin in the cortical part of the cytoskeleton (Liu et al., 1987), and both form triangulated hexagons beneath the cell membrane (Figure 6.5) (Weinbaum et al., 2003). They also link short proto-filaments of actin to the centre of a suspension complex, at least in erythrocytes (de Oliveira et al., 2010), with the spectrin fibres under tension and proto-filaments under compression (Figure 6.6) in a manner fairly similar to the spokes and hub of the tensegrity bicycle wheel described previously (Figure 1.9, page 8). There are about 33,000 of these suspension units and they are considered as the parts of

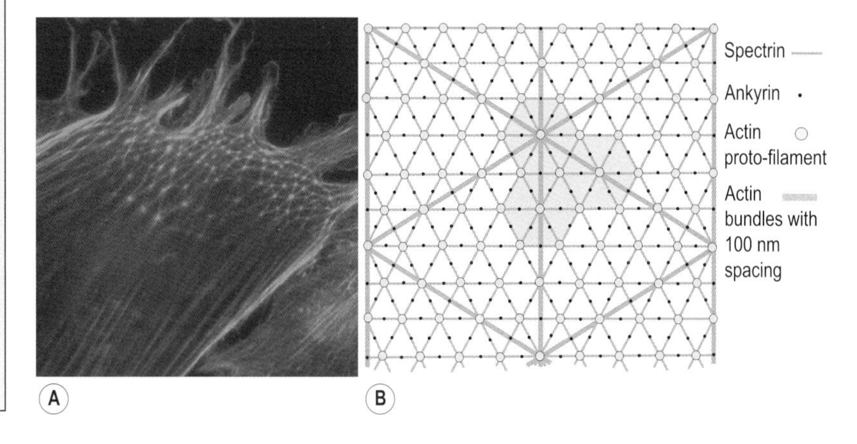

Figure 6.5
A Geodesic forms in the tensed cortical cytoskeleton of a living human cell. (reproduced courtesy of © Donald E. Ingber, Scholarpedia); B schematic diagram showing an idealized relationship (hexagons) between bundles of actin filaments, actin proto-filaments and spectrin filaments fixed to ankyrin in the lipid membrane of an endothelial cell (not to scale) (re-drawn with modifications from Weinbaum et al., 2003).

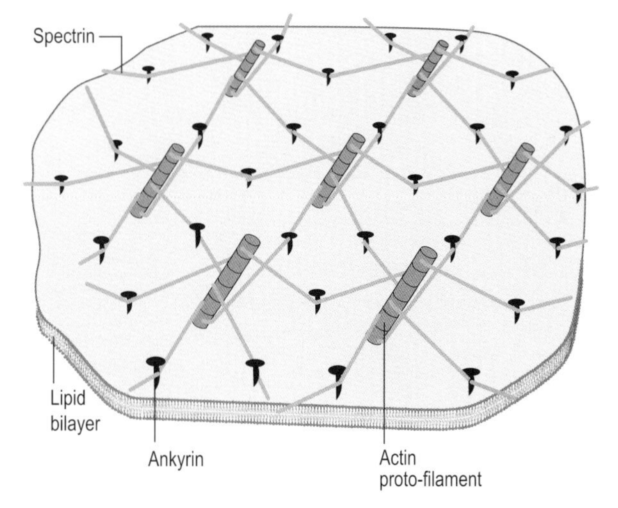

Figure 6.6
Schematic drawing of actin proto-filaments suspended within a triangulated network of spectrin filaments fixed to ankyrin in the lipid membrane; each actin proto-filament can be compared with the hub of the tensegrity bicycle wheel in Figure 1.9 (re-drawn with modifications from Zhu et. al., 2007).

a high-frequency icosahedron that encloses the cell and links to the internal cytoskeleton, with about 85% appearing as hexagons, 3% as pentagons and 8% heptagons, confirming nature's predilection for the hexagon (Figures 2.3 and 2.4, page 14) (Zhu et al., 2007).

Collagen

Collagens, of which there are about thirty different types, consist of even more complex structural hierarchies, and as the most widespread structural molecule in humans are one of the main distributors of tissue tension (Fratzl, 2008) (Figure 6.7).

In collagen type I, repeating sequences of amino acids spontaneously form a left-handed helix of procollagen, with three of these combining to form a right-handed tropocollagen molecule. These then pack together in a staggered array to form microfibrils, fibrils and fascicles within bone, tendons, ligaments and fascia, etc. Collagens also have complex links with other molecules such as proteoglycans and elastin, and influence the properties of tissues at higher scales (Cavalcante et al., 2005).

Collagen is further interesting from a geometric perspective because each left-handed procollagen closely matches the sub-helix in Figure 6.2b. Three of these then join together to enclose the central core of the tropocollagen molecule. In addition, every third amino acid in each of the procollagen chains is a glycine that has a hydrogen atom projecting into the central core of the tropocollagen molecule and because the procollagen chains are staggered, these particular hydrogen atoms coincide *precisely* with the vertices of a right-handed tetrahelix (Sadoc & River, 2000; Lord, 2002) (Figure 6.8).

While the tetrahelix is the geometric template for simple biological helixes, so combining left and right-handed chiralities makes them much more stable, and

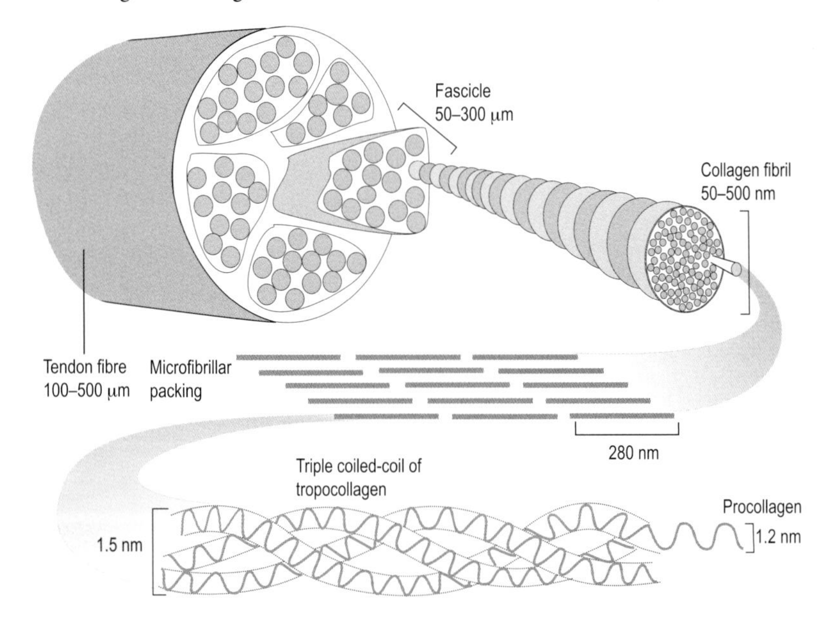

Figure 6.7
Schematic diagram of a collagen hierarchy within tendon (re-drawn with modifications from © Scarr, 2010, Elsevier).

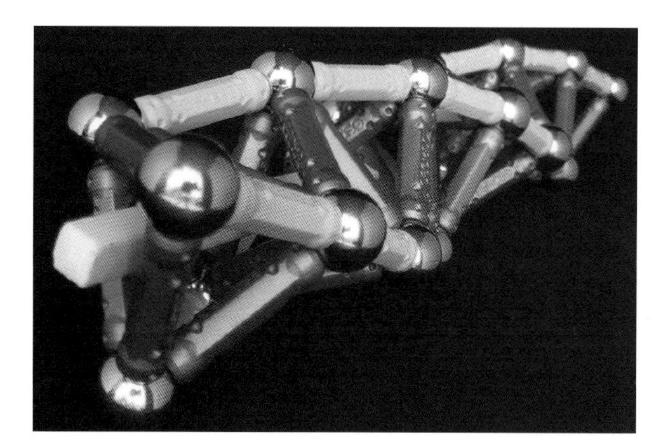

Figure 6.8
A representation of the glycine hydrogen atoms (spheres) of procollagen that project inwards towards the tetrahelical core of the tropocollagen molecule (the square rod indicates the centre of the core).

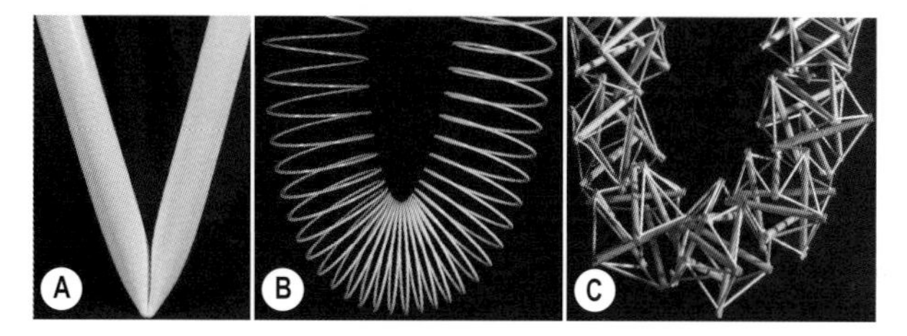

Figure 6.9
The material properties of a tube wall affect its ability to bend: A a rigid tube buckles and restricts fluid flow; but B a coiled spring and C crossed-helical tensegrity (T-icosahedral chain) remain open. © Rory James

Levin emphasised the significance of this with a chain of tensegrity icosahedra, i.e. a crossed-helical tube (Levin, 1986, 2002) (Figure 6.9b, above).

The helical tube

One of the functions of collagen is in the helical reinforcement of flexible tubes, and the walls of the gastro-intestinal tract (Carey, 1920a; Gabella, 1987), respiratory tract (Carey, 1920b) and arteries (Holzapfel, 2008), etc. all contain concentric layers of collagen fibres arranged in alternating left and right-handed helixes. Crossed-helical fibre arrays provide reinforcement, which then allows elastic tubes to bend smoothly without buckling and react to changes in volume in particular ways – particularly useful properties for maintaining fluid flow (Figure 6.9). An optimum fibre angle of 54.44° will balance both longitudinal and circumferential stresses, which means that a lower angle will resist being stretched in length (Figure 6.10a) and a higher angle resist increases in diameter (Clark & Cowie, 1958; Shadwick, 2008) (Figure 6.10b). All the tubes described above are reinforced by collagen fibres with a similar angle (note that the term 'fibre angle' is in relation to the tube axis and differs from the standard engineering use of 'helical angle' that is perpendicular to it) (Figure 6.10c).

Within the heart, muscle and collagen fibres show a gradual change in orientation between inner and outer walls, from ~55° in one direction to ~55° in the other, and tangential spiralling in a transverse plane (Purslow, 2008), and these alignments form a helical coil of muscle overall that contracts with left- and right-handed twisting motions (Buckberg, 2002). As the heart typically produces a left ventricular ejection fraction of 60% for a muscular contraction of just 15%, this confirms the mechanical efficiency of this helical configuration.

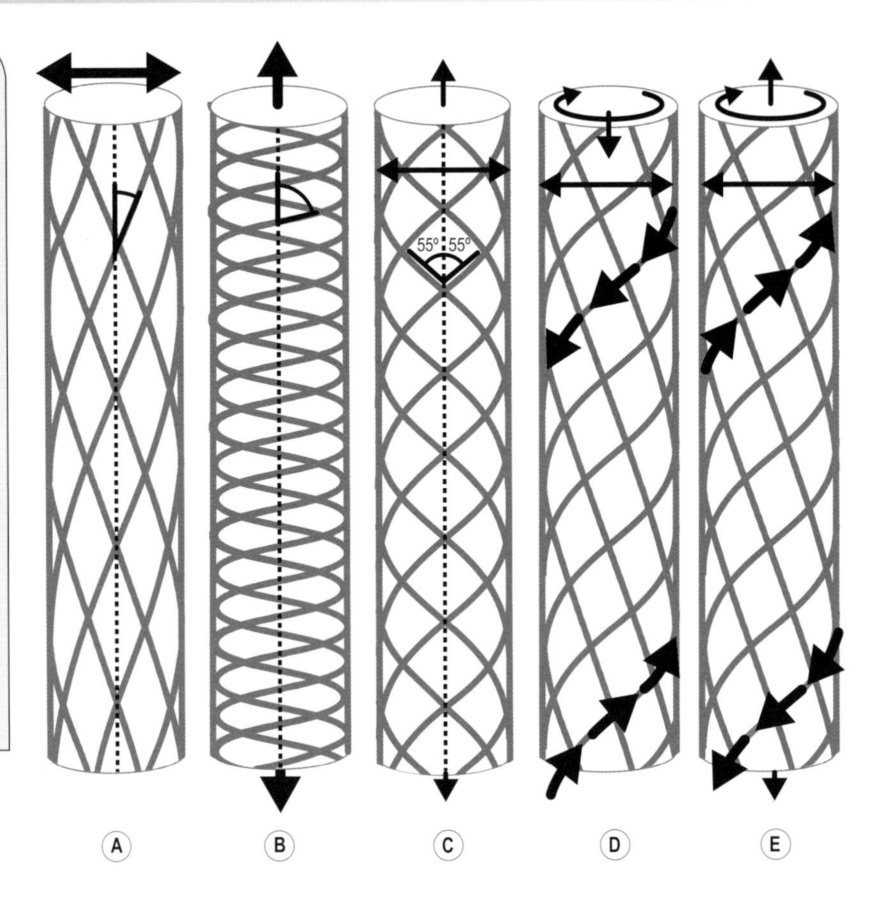

Figure 6.10
Diagram to show the influence of crossed-fibre angles on the response of a helical tube to increases in volume: A a low fibre angle increases tube diameter; B a high fibre angle increases tube length; C the optimum fibre angle of 54.44° balances both influences; D left- and right-handed helixes with different angles become shorter and introduce torsion; E but a tensegrity crossed-helix will actually *increase* in length and diameter (a 'negative' Poisson ratio) (re-drawn with modifications from Scarr, 2013).

Levin also described the heart as a tensegrity pump based on the expansion and contraction of the tensegrity-icosahedron (Figure 3.4, page 28) and jitterbug (Levin, 2002) (Figure 2.18, page 22); and because anatomical structures and tensegrity models are similar in that they both consist of many discrete parts, a tensegrity chain then serves as the ideal model for biological crossed-helixes.

Essentially, these helical arrangements form tubes, from the outer walls of bacteria (Jones et al., 2001) and plant cells (Lloyd & Chan, 2002) to arteries (Holzapfel, 2008) and the embryonic notochord (Koehl et al., 2000), and they are all made from multiple components integrated into a mechanical unit; and this similarity with tensegrity then means that their dynamics can be easily modelled and more complex hierarchical structures explained (Fuller, 1975, 750.00; Flemons, 2012).

Tubes within tubes within tubes...

The myofascial tube

Fascia is a collagen-rich connective tissue with considerable influence over the transmission of tension (Passerieux et al., 2007; Stecco et al., 2009; Schleip et al., 2012) and connects every bone and muscle into a single tensioned network. It naturally develops into compartments, or 'tubes within tubes', that are

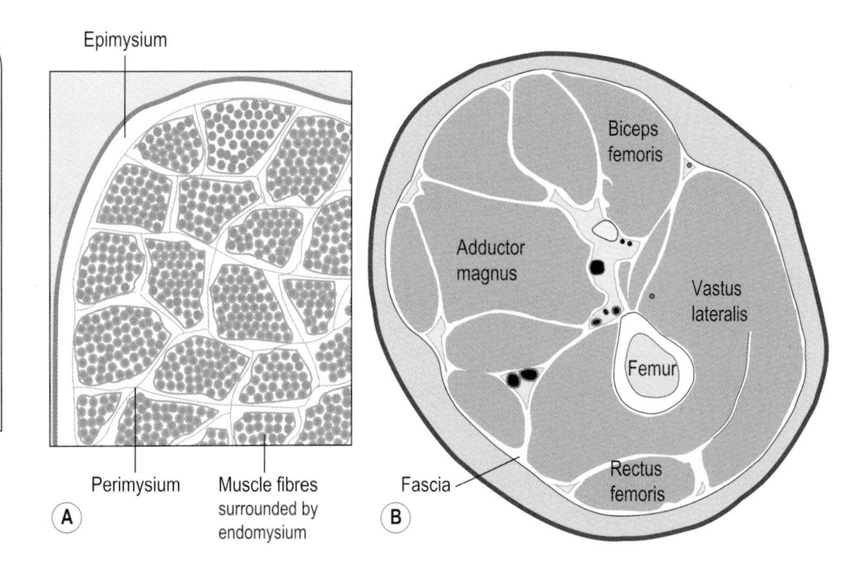

Figure 6.11
A Schematic diagram of a transverse section of muscle showing the hierarchical arrangement of myofascial tubes (not to scale); B transverse section of the thigh showing larger myofascial tubes surrounding individual muscles and the entire limb.

particularly noticeable in cross-sections of the limbs (Figure 6.11). The myofascia is a specialisation that is intimately connected with muscle and, as a delicate network of endomysium surrounding individual muscle fibres, it is continuous with the perimysium ensheathing groups of fibres in parallel bundles (fasciculi), and epimysium covering the entire muscle and fascia investing whole muscle groups (Turrina et al., 2013). All these sheaths coalesce and contribute to the transmission of the force generated within muscle fibres to tendons and inter/extra-muscular fascial attachments (Huijing & Baan, 2003).

Although there is a lot of variation in the alignment of fibres in these tissues, particularly in the endomysium, the peri- and epimysium are reinforced by two helical crossed-ply sets of collagen with an average resting fibre orientation close to the 'ideal' 55° (Purslow 2008; Benetazzo et al., 2011) (Figure 6.10c).

The body wall

Similar helical tubes with alternating layers of collagen fibres close to 55° (Figure 6.10c) also surround the body walls of a huge diversity of species, from worms (Shadwick, 2008), eels (Hebrank, 1980) and reptiles (Figure 6.13a), to fish, dolphins, whales (Pabst, 2000), cows (Purslow, 2008) and even humans (Benetazzo et al., 2013) (Figure 6.12). They have also been described in the tongues of mammals and lizards, the arms and tentacles of cephalopods, and the trunks of elephants (Kier & Smith, 1985). Even the gross arrangement of thoraco-lumbar and abdominal musculature in humans has a spiral appearance, if only in part (Figure 6.12), and the functional significance of this to human movement has not gone unnoticed (Stecco, 2004; Levin & Martin, 2012; Myers, 2014).

A more fundamental kind of geometry

A particularly interesting example is found in the limb scales of the Pangolin *(Manus spp)*, where the left- and right-handed helixes are arranged at different angles (Figure 6.13b), and *may* be a geometric reflection of collagen alignment in

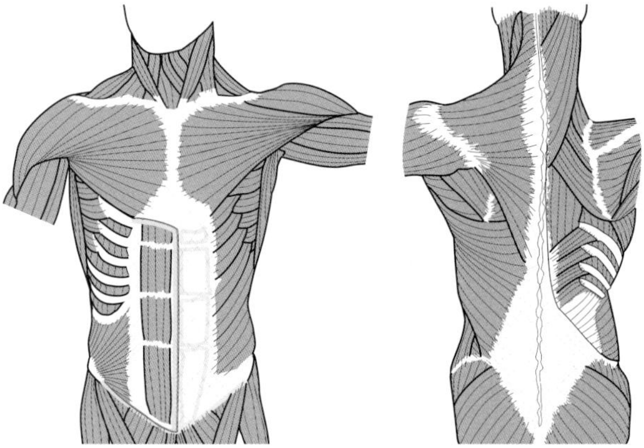

Figure 6.12
The muscles of the trunk form partial spirals around the body that continue into the neck and limbs.

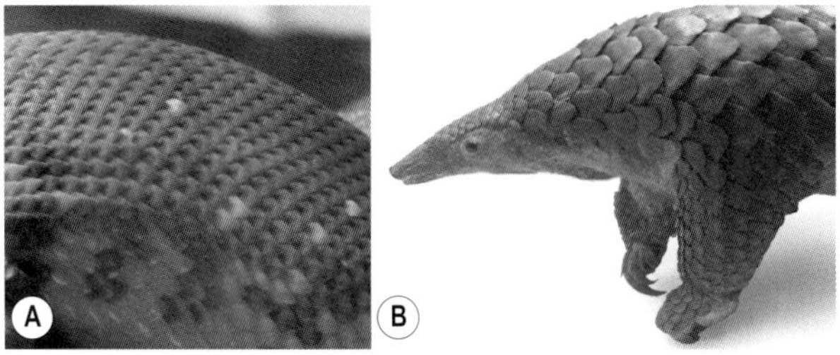

Figure 6.13
The crossed-helical pattern of scales on a: A *Boa constrictor*; B pangolin *(Manis javanica)* (reproduced from © Scarr, 2013, Elsevier).

the myofascia and skin of all mammals (Scarr, 2013) (Figure 6.10d); but we will have to wait until Chapter 10 to find out why this might be case.

Stirring the pot

Biological helixes appear in so many different situations because they are all subject to the same rules of geodesic geometry, close-packing and minimal-energy that formed the tetrahelix in the first place, and it is largely the geometric constraints of developing cylinders and tubes that cause their dynamic contents to settle into this particular arrangement (Pickett et al., 2000; Nisoli et al., 2010). In addition, their structural properties are similar to the stick-and-string tensegrity models in that they are composed of multiple distinct components; they are intrinsically tensioned and hierarchical; their dynamics are non-linear (Figure 4.1b, page 34); they allow changes in shape without buckling or collapsing and they directly influence the motion of even more complex structures (Levin, 1986, 2006); and after all, it is movement that enables biological organisms to express life.

7 The ease of motion

Biotensegrity is an efficient, low-energy mechanical system that when applied to complex biological structures, overcomes many of the inconsistencies that were present in the old view of biomechanics. Stephen M. Levin (2014)

Andreas Vesalius (1514–1564), professor of anatomy and surgery at Padua University in Italy, was dissatisfied with the dominant but erroneous anatomical wisdom that had prevailed since the time of Galen (129–200 CE), some fourteen hundred years previously, and refreshingly performed his own dissections (Rifkin & Ackerman, 2006, p. 14). What he found was that human anatomy is actually very different to that which had been previously described and eventually wrote his most influential book *De Humani Corporis Fabrica* (Figure 7.1a). The mathematician Giovanni Borelli (1608–1679) then compared the anatomy of human movement with man-made machines of the day (Figure 7.1b) and wrote *De Motu Animalium* (Ethier & Simmons, 2007, p. 6), setting the benchmark for orthodox biomechanics for the next three hundred years.

Borrelli's biomechanics dominated, much like Galen's ideas did, because no one had come up with a better explanation for what could be observed. The bones of the skeleton were considered to stack on top of one another like a pile of bricks (Figure 1.5, page 6) and resist gravity through a complicated system of levers, because that was the well-known building system common to man-made structures; but as we have already seen, their value in describing biological movement is highly questionable (Levin, 2006).

A feature of vertebrates is their ability to move through the use of synovial joints, and biomechanics has traditionally considered such movement through a

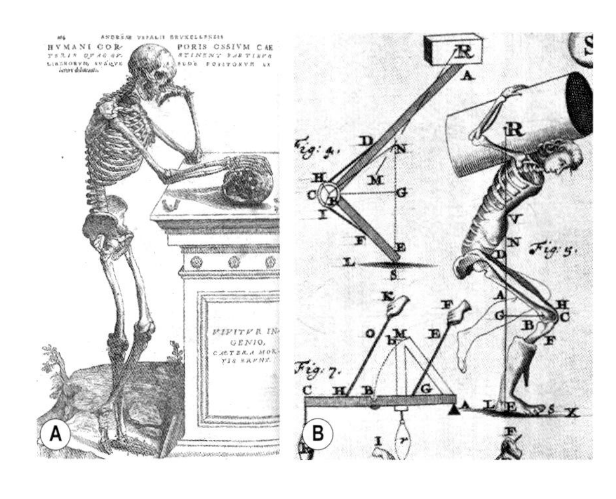

Figure 7.1
A An illustration from *De Humani Corpora Fabrica* by Vesalius (1543); B An illustration from Borelli's *De Motu Animalium* (1680) showing human movement based on a mechanical lever system.

system of free-body analysis, where bones and joints are isolated and simplified down to some arbitrary 'minimum' (Figure 4.3, page 37) (Standring, 2005, p. 134). In this scenario, bones compress each other and move around a fixed axis, connective tissues are rarely taken into account and the continuity of force transfer between joints is usually ignored. However, this system is incomplete, far too simplistic and no longer fits with modern biology.

Replacing the old with the new

Living bones move around complex *helical* axes (Standring, 2005, p. 110) and it is unlikely that they compress each other much during normal movement (Levin, 2002; Andrews et al., 2011). However, because the amount of force produced by individual muscles varies according to their length, load and the joint angle, they must be part of a much more complex movement system.

Dinosaurs and the Forth Bridge

As an orthopaedic surgeon, Levin observed many normal and not-so-normal joints, but he was disillusioned about the wisdom of the Borelli model (Levin, 1981). He felt that the conventional stack-of-blocks image used to describe the spine did not explain the complexity of its structure and motion. In 1975 he began studying the skeletons of dinosaurs and mammals at the Smithsonian Museum of Natural History in Washington, D.C. and over the course of several months, spent many hours consulting with curators and paleontologists (Levin, 2014) (Figure 7.2a).

Sitting on a bench outside the museum, he pondered what he had observed: how could the immense weight of these creatures have been supported by a lever system that would have placed enormous stresses on their tissues (Figure 4.2, page 34). The conventional view was that their bodies were cantilevers similar to the Forth Bridge (Figure 7.2b), an idea that had been around for decades (Thompson, 1961, p. 245), but this had none of the motion of a living organism.

Snelson's tower

Across the mall from where Levin was sitting was a sculpture by Kenneth Snelson; the sixty-foot *Needle Tower* (Figure 7.3). This structure appeared to be very strong yet was light and flexible – the sort of features that would have

Figure 7.2
A Skeleton of *Diplodocus longus* (Smithsonian Museum, Washington D.C.) B a model of the Forth Railway Bridge in Scotland (London Science Museum).

developed through evolution in a living organism, particularly in those long and heavy dinosaur necks (Figure 7.2a). Snelson's sculpture then stimulated the idea that tensegrity might be a viable alternative to understanding biological structure, and rather than the dorsal neck muscles acting like the jib of a crane (lever) in holding up the immense weight of the head, a tensegrity configuration would distribute this force *throughout* the neck and avoid stress concentrations and points of potential weakness (Figure 7.4). During development, tensional

Figure 7.3
Kenneth Snelson's *Needle Tower* 1968, aluminum and stainless steel; Hirschorn Museum, Washington D.C.

Figure 7.4
A chain of tensegrity-icosahedra supports itself and models the vertebral column of long-necked animals, etc.

forces would remodel the bony contours according to Wolff's Law, etc., and modify the positions and orientations of soft-tissue attachments that contribute to the apparent complexity of shapes. While such a mechanism would be a distinct advantage in long-necked animals, why would it stop there (Levin, 2002)?

The vertebrate spine

Traditionally, the erect spinal column of humans has been viewed as a tower of solid blocks and squidgy discs that transfer their compressive load down through each segment to the pelvis (Figure 1.5, page 6), but a vertical spine is a relative rarity amongst vertebrates. Most other animals have little or no use for a compressive vertebral column, which is frequently portrayed as a horizontal truss (Gordon, 1978, p. 234) and cantilever support system (Figure 7.2b) and, as the main difference in vertebrate anatomies is in the detail, it seems reasonable to suppose that they all have some structural properties in common.

The model in Figure 7.5 shows two human vertebrae held apart by a single ring of tension, i.e. it is a simple tensegrity configuration that is similar in principle to Snelson's *Early X-piece* (Figure 1.1, page 1) and Flemons' spinal tensegrity model (Flemons, 2007) (Figure 3.10, page 31). If the huge complexity of soft tissues surrounding bones act in the same way *in vivo*, they would enable the spine to

Figure 7.5
Two mid-thoracic human vertebrae held *apart* by a continuous ring of tension and the particular alignment of bones (struts).

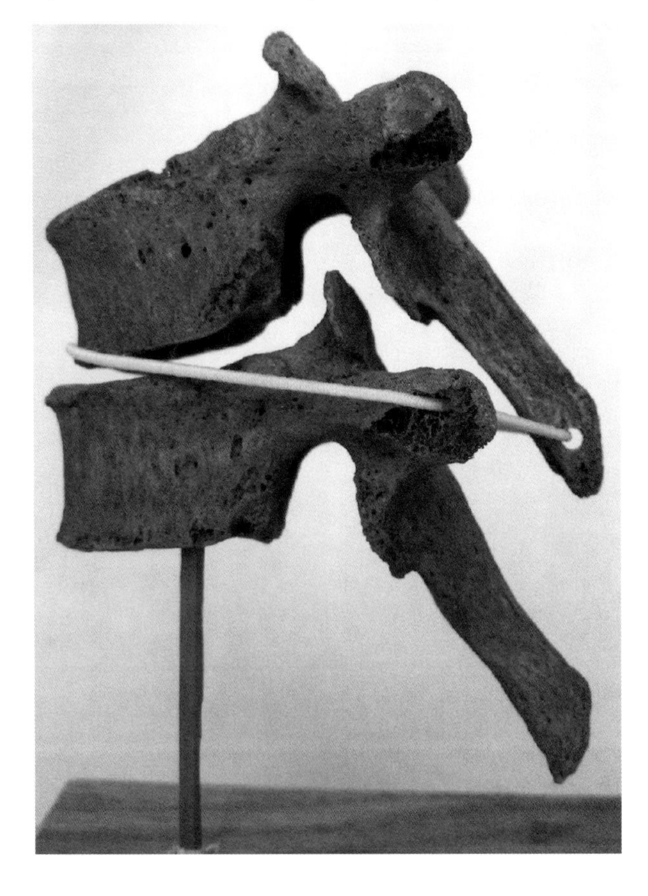

function equally well in any position, irrespective of the direction of gravity or loading, and with the *minimum* of compression (Robbie, 1977; Levin, 2002).

Intervertebral discs (with their helical reinforcement) can then be considered as facilitators of movement, with compressive forces dissipated largely through a tensioned system that avoids localised stress concentrations. In this example, collagen fibres are arranged in concentric lamellae with opposing orientations in alternate helical layers of 65°, a higher angle than most other helical 'tubes' (~55°) (Figure 6.10) that provides tensile reinforcement during bending and twisting, and stabilises the disc against bulging (Hukins & Meakin, 2000). Future research that considers the spine and other joints from a biotensegrity perspective is likely to lead to a better understanding of spinal mechanics.

The biotensegrity joint

During surgery, Levin correlated his biotensegrity thesis with observations at the operating table and found that tightening up the cruciate ligaments in a knee joint caused the bones to move apart, something that was inconceivable in a lever system (Levin, 2014). He also saw that normal joints always had a slight spacing between the bones and that this could be compromised by pathology, but there was no known mechanism that could make this space possible; it was like the bones were 'floating' in the soft tissues (Terayama et al., 1980; Levin, 1981).

The wheel

Using the analogy of the tensegrity bicycle wheel, where the tensioned spokes hold the central hub and outer rim in perfect balance (Figure 1.9, page 8), Levin suggested that the scapula functions as the hub of such a wheel and is, in effect, a sesamoid bone (see footnote) that transfers its load to the axial skeleton through tensioned muscular and fascial attachments (Levin, 1997, 2006). In addition, the sterno-clavicular joint is not really in a position to accept much compressional load, and the transfer of axial compression across the gleno-humeral joint could really only occur when loaded at 90° abduction. As synovial joints are essentially frictionless inclined planes, they would rely heavily on ligamentous and muscular tension in all other positions just to remain stable. The enormous stresses generated within the tissues are then likely to exceed their strength capability, and be very expensive in terms of the amount of energy needed to sustain even the simplest of movements, thus making this an impossible situation (Levin, 1997, 2006).[1]

However, if the humeral head is compared to a bicycle wheel hub (Figure 1.9) and suspended from the glenoid 'rim' by a tensegrity configuration of muscles, ligaments and other connective tissues, the shoulder would be able to function equally well with the arm in any position – within the physiological constraints of the tissues of course (Figure 7.6).

---

[1] Sesamoid bones, named by Galen because of their resemblance to sesame seeds, are usually embedded in tendons or other dense connective tissues, e.g., the patella bone and sesamoids beneath the ball of the foot. Their functional role remained obscure before Levin's biotensegrity model (Standring, 2005, p. 1523).

Figure 7.6
Acrobatic gymnastics clearly shows that the entire body weight can be supported by tension across the gleno-humeral joints. Gymnasts Éleonore Lachaud and Clara Guinard. Reproduced from © Trampoline club du Dauphiné, Wikipedia.

In a similar way, the tibia can be likened to a wheel hub within the femoral rim of soft-tissue attachments, just like the shoulder where load bearing across the joint is significantly tensional and compressional forces are redistributed through the tensioned network. The pelvis is also like a wheel with the iliac crests, pubis and ischia representing the outer rim, and the sacrum (compared to the hub) tied in with strong sacro-iliac, sacro-tuberous and sacro-spinous ligaments, etc. The femoral heads are like hubs within the 'spokes' of the ilio-femoral, pubo-femoral and ischio-femoral ligaments (Levin, 2007; Flemons, 2007).

Floating sesamoids

These descriptions are obviously gross simplifications because there are numerous tensioned structures that cross the joints but they illustrate the 'floating bones' principle. The ability of bones to move in relation to each other is an essential feature of vertebrate anatomies, and it is likely that their joints would all operate within the same mechanical framework. As lever systems, they would be prone to bending moments and damaging stress concentrations

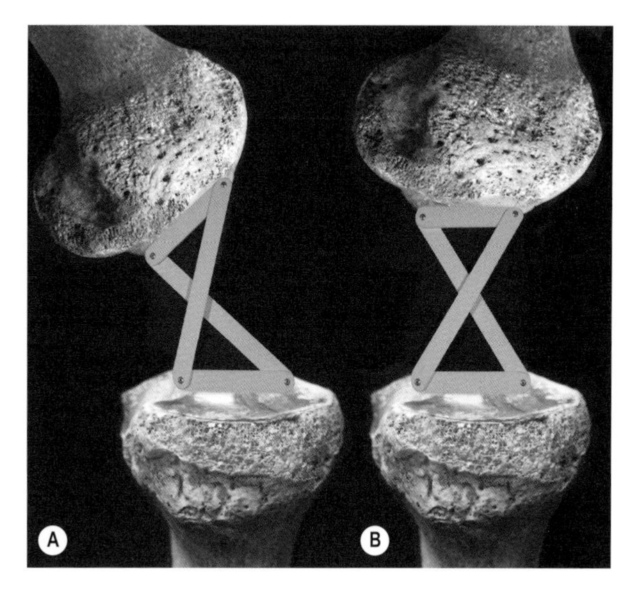

Figure 7.7
Lateral view of the knee with the cruciate ligaments modelled as a crossed 4-bar kinematic chain that guides movement. A Flexion; and B Extension (Levin, 2007)

(Figure 4.5, page 38), and would require the control of a much more complex nervous system than is currently evident (Turvey, 2007). As closed-chain kinematic systems, however, the physical constraints of the tensioned structural network itself would guide motion (Figure 7.7) and, as tensegrity configurations, all the tissues surrounding the joint would be integrated into a functional unit that distributes stresses harmlessly throughout the system and becomes stronger as it is loaded (Figure 3.4, page 28). The tensegrity-icosahedron then becomes the ideal model for understanding joint motion because it displays all the essential characteristics necessary for movement (Levin, 2006).

A little bit of space

One of the features of the tensegrity model is that compression is discontinuous, i.e. it is not transferred directly between the struts (bones), and examination of the knee and other joints has shown that there is little or no compression between normal joint surfaces *in vivo*, even when they are pushed together (Terayama et al., 1980; Levin, 1981)!

The knee

While Levin regularly observed that there was a small space between the bones, he found that his own knees also required some surgical intervention (nothing major, just a little arthroscopic tidying up) and took the opportunity to test some of these ideas on himself. Fully conscious and watching the video in real-time, he requested that the operating surgeon perform some particular manoeuvers that might verify the 'floating bone' principle of biotensegrity. As this added trivially to the operating time and nothing to the risks involved, the surgeon agreed. What was then observed was a continuous spacing of 1–3 mm between both the patella-femoral and femoro-meniscal articulations, even when the quadriceps were actively contracting, or the sole of the foot was pushed towards the hips and simulated weight bearing (Levin, 2005) (Figure 7.8).

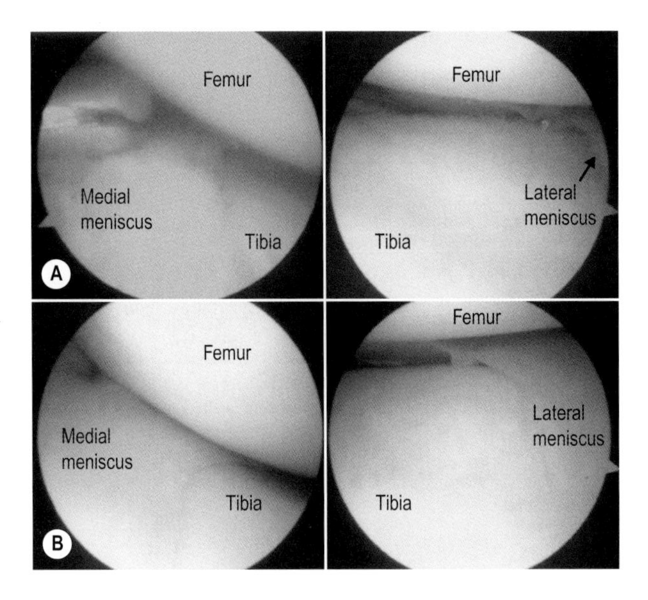

Figure 7.8
Anterior views of the left knee joint showing the constant spacing between the femur and menisci when: A unloaded; B loaded. Note instrument in space between the femur and medial meniscus (top left). Reproduced courtesy of © Stephen Levin (2005)

This spacing between two supposedly 'compressed' bones inevitably suggests that something a little bit different from the standard biomechanical model must be going on. Although the details of the soft tissue mechanics that could lift the femur into the 'air' are lacking, there are no obvious levers, and biotensegrity would seem to be the only possible explanation for these findings. Although significant pressures have been observed across other joint surfaces *in vivo*, it is likely that the substantial thickness of the sensors placed between the bones may have adversely influenced the results in these instances (Rikli et al., 2007). In any case, a review of the literature casts doubt on the ability of the menisci to act as shock absorbers (Andrews, 2011).

In addition, it should be recognised that soft tissues are also hierarchical structures in their own right (Fratzl, 2008; Qin, 2009) (Figures 3.8, page 30 and 6.7, page 52), and that crossed-helical myofascial networks may also be maintaining some axial compression that contributes to the joint spacing (Flemons, 2012). We will see later that the difference between a 'cable' and 'strut' is really just a matter of perspective in a biotensegrity context.

Sliding surfaces

Synovial joint surfaces are incongruous, even with cartilage between them, and the amount of surface contact in cadavers varies with joint position and load because of this (Eckstein et al., 1995a). While a cyclic loading during movement *in vivo* would actually improve chondrocyte activity and cartilage lubrication (and significantly reduce peak surface pressures in any case) (Eckstein et al., 1995b), the ability of joint surfaces to slide over each other without transferring major compressional forces is still compatible with tensegrity in a biological context (Fuller, 1975, 715.01; Levin, 2002). It is thus probable that normal cartilage functions more as a protective end-cushion rather than major load bearer, with tensioned muscles and connective tissues regulating the joint spacing and surface pressures.

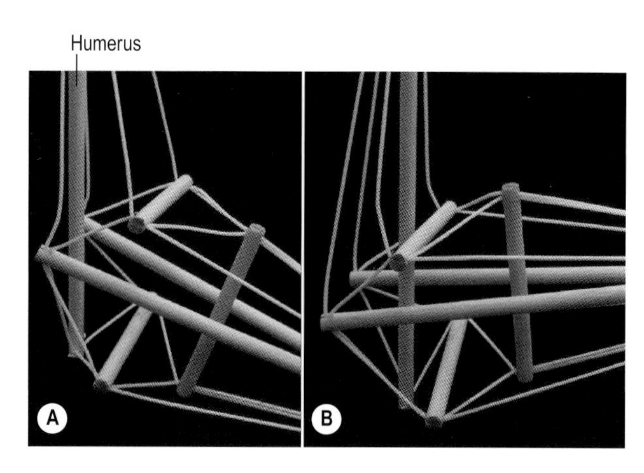

Figure 7.9
A modified T-icosahedron model of the elbow in: A flexion; and B extension; with some of the struts elongated to correspond with the bones and the soft tissues controlling motion.

A note of caution

This issue of joint compression is not really something that mechanical engineers have had to be overly concerned about and, as a consequence, have introduced a particular tensegrity classification that may be useful from an engineering perspective but that is definitely *not* transferable to biology: Class 1, where the struts don't transfer compression between themselves, as defined by Snelson (Heartney, 2009, p. 20) (Figure 1.1, page 1) and Fuller (1975, 700.011); and classes 2, 3, 4, etc. (Skelton & de Oliveira, 2008, p. 8), where different numbers of struts compress each other and become part of lever systems.

Firstly, we should note that this particular classification is an engineering convenience and not relevant to living organisms; secondly, we have already seen how the application of levers to biological motion is erroneous and misleading. Man-made structures are constrained by the properties of the materials used, practicality of construction and cost (and in a sense anything goes), while biology is constrained only by the laws of physics with different patterns and shapes naturally developing because they are the most energy-efficient configurations. The inevitable generation of bending moments and shear stresses within a class 2, 3, 4, etc. system could simply not be sustained as part of normal biological function (Levin, 2002) (Figure 4.5, page 38), and such descriptions are misleading because they disregard the hierarchical nature of this region.

While the general principles of this biotensegrity structural system might be starting to become clear, the precise anatomical details that support the findings described above are still being worked out, and a closer examination of the elbow shows that it is much more interesting than a lever (Figure 7.9).

A bit more detail

The elbow

The elbow is a joint that is commonly described as a uniaxial hinge that permits movement in a single plane and the pivot of proximal forearm rotation. It enables the length of the upper limb to vary through flexion and facilitate the hand in carrying out its functions of grasping and manipulation. Although it is well established that the humerus and radius do *not* make contact during movement (even though muscles are pulling them together) (Kim et al., 2002), the main function of some of these muscles was poorly understood until recently. The biotensegrity model now gives them a respectable function (Scarr, 2012).

Something rather peculiar is going on

The brachioradialis (Figure 7.10) is a particularly odd muscle because of its minimal activity during slow easy flexion, even though its relatively high attachment on the supracondylar ridge gives it a mechanical advantage over the other flexors. It also differs from them in that its muscle bulk is mostly distal to the joint, which must increase inertia of the limb and require more effort to move it during rapid movement, yet brachioradialis is mainly active in rapid and forced flexion *and* extension (Standring, 2005, p. 879). Bizarrely, it is supplied by the *same* nerve (radial) that innervates the extensor muscles, triceps and anconeus, its supposed antagonists!

It is well known that the elbow flexors are not a single functioning unit (Pauly et al., 2005), and as an 'associate flexor', the distal end of brachioradialis is sometimes cut and repositioned to replace the loss of other muscles after injury, which shows how little its 'normal' function is generally regarded. Another muscle that is also poorly understood is anconeus (Standring, 2005, p. 881), which is active throughout all movements of the elbow and always precedes triceps (Le Bozec et al., 1980), and uniquely is the only elbow muscle that is not directly connected to the overlying fascia. Rather vaguely, it has been considered as a 'stabilising' muscle, and all these factors suggest that the principle functions of brachioradialis and anconeus may be somewhat different.

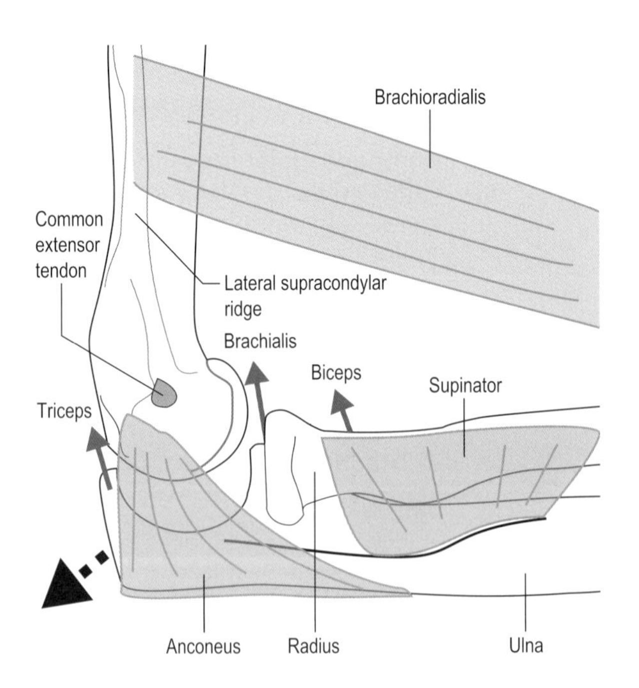

Figure 7.10
Lateral view of the right elbow showing the direction of presumed distraction force exerted by brachioradialis (black dotted arrow); the humero-ulnar joint spacing has been increased for clarity.

A respectable function

The bulk of brachioradialis is clear to see on the outer part of the forearm and its other insertion is on the distal radius (near the wrist) and, as a result, acts in a trans-articular or shunt capacity that indirectly pulls the ulna towards the humerus during flexion (Williams & Warwick, 1980, p. 579) (Figure 7.10). However, because of its high attachment on the lateral supracondylar ridge, brachioradialis must also create a distraction force between the humerus and ulna, which would then be contained by triceps, anconeus, the collateral ligaments and associated fascia, etc. Both brachioradialis and anconeus muscles acting together could then contribute to the joint surfaces gliding, or even floating alongside each other without necessarily being compressed, which makes this analysis compatible with the elbow as a tensegrity configuration (Scarr, 2012).

The well recognised coupling of all these muscles to the posterior part of the capsule through the *same* nerve supply (radial) then starts to make sense, because they would be able to quickly receive proprioceptive information from the capsule and other tissues, and become part of a rapid response unit that regulates pressure/spacing across the joint surfaces during movement (Scarr, 2012) (Figure 7.11). As cables in a closed-chain kinematic tensegrity configuration, these muscles also become part of the global tension network

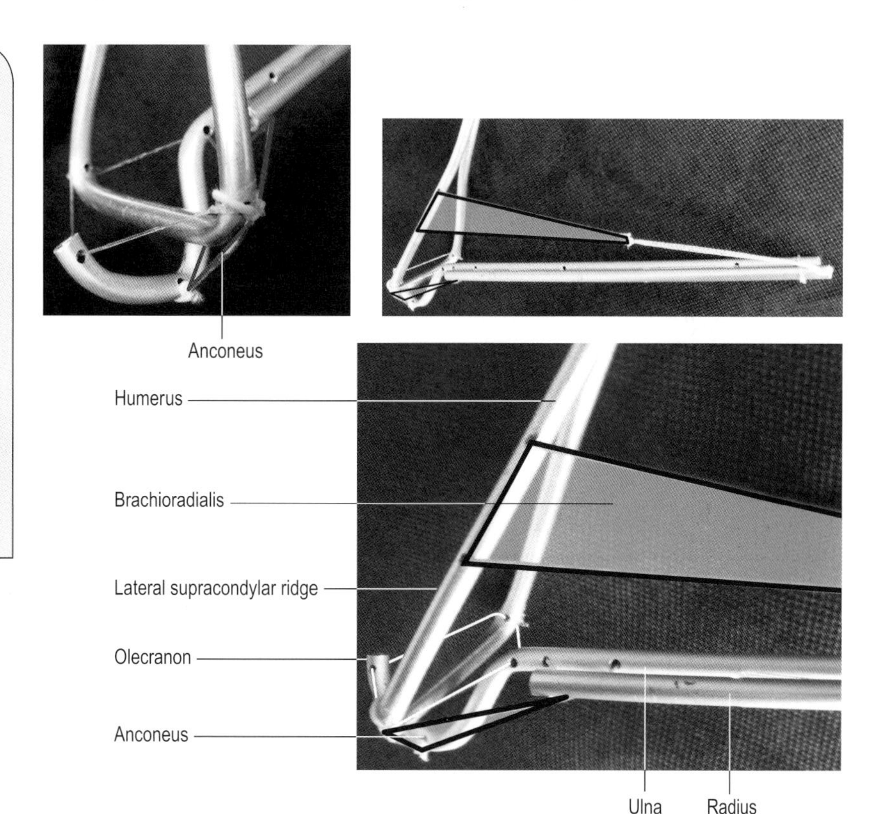

Figure 7.11
A simple tensegrity model elbow demonstrates the 'floating bone' principle (rather than anatomical accuracy). Antero-lateral view showing bones tensioned by connective tissues, brachioradialis and anconeus muscles. Insets: top left – postero-lateral view; top-right – overall view. Reproduced with modifications from © Scarr, 2012, Elsevier.

that controls movement. The helical trajectory of joint motion then relates to the tensegrity-icosahedron (Figure 3.4, page 28 and Figure 7.9) and energetic twist of the jitterbug (Figures 2.18 and 2.19, pages 22 and 23), and a simple mathematical analysis of the elbow based on the T3-prism (Figure 3.1, page 25) also supports the tensegrity principle (Cretu, 2009).

A little resumé

Biotensegrity is a complete structural system and it could be said that these apparently isolated examples have just been thrown in to justify the cause, but this is not really a fair assessment on its own. Although it is true that much of the evidence at the macro-scale is rather circumstantial, the concept is best considered in its entirety before coming to a conclusion.

Biotensegrity explains things that traditional biomechanics has struggled with, or simply ignored, and presents tangible models that help to answer some of those difficult questions. The bicycle wheel model explains how synovial joints can function without overstressing the soft tissues surrounding them, even under very high loads, and how the hub-like bones can 'float' within them. It even gives sesamoid bones a proper function for the first time! An examination of the knee *in vivo* reveals that the femur must be 'floating' above the tibia – and biotensegrity is currently the only explanation that could make this possible – even though we don't know the precise details as yet. A consideration of the elbow also explains how particular muscles and other tissues could be directly responsible for holding joint surfaces apart, and regulating their motion. The vertebrae are then considered to behave as part of what is essentially a tensioned structure, that enables the spinal columns of humans and other animals to function equally well in both horizontal and vertical positions.

Biotensegrity explains how the bones on either side of a joint can remain stable yet move with the minimum of effort, and how the soft-tissues are able to guide them. It describes how each tissue can be integrated into a complex tensional network that extends in all directions, and forms an 'automatic shifting suspension complex' that couples multiple joints (and the entire body) into a functioning unit.

The principles of biotensegrity explain how an acrobat can balance on one hand (Figure 7.6), and a weightlifter can lift objects that are 'too heavy' for the soft tissues to support (Levin, 2006) (Figure 4.4, page 37). Biotensegrity principles also recognise that the basic geometries of the helix and icosahedron are embedded within the structural hierarchies that allow them to do this.

While the tensegrity-icosahedron is a model that contains all the hard struts on the inside of the structure, and together they form an endo-skeleton (like the bones mentioned above), an example of a reverse situation occurs in the skull where bones surrounding the brain essentially form a hard exo-skeleton with significant tensional tissues occurring on the inside. The next chapter takes the simple tensegrity-icosahedron and bicycle wheel models and uses them to examine some particular aspects of cranial development and intricacies of the avian lung.

8 The hard and the soft

...ideal geometries...pervade organic form because natural law favours such simplicity as an optimal representation of forces. Stephen Jay Gould (Thompson, 1961, xi)

The cranial vault and the bird's lung are two distinct anatomical structures with quite different functions, but they share similarities in their construction, i.e. their underlying tensegrity mechanics are the same but different models are best used to describe them. Until quite recently, many aspects of cranial development have been poorly understood, and the exceptional mechanical strength of the avian lung has also presented a bit of a puzzle. This chapter shows how the biotensegrity model helps to explain some aspects of these structures and improves our understanding of their functions.

The cranial vault

The cranial vault is a set of bones that surround the top, back and sides of the skull, i.e. those that enclose and protect the brain, with bones of the cranial base lying underneath and

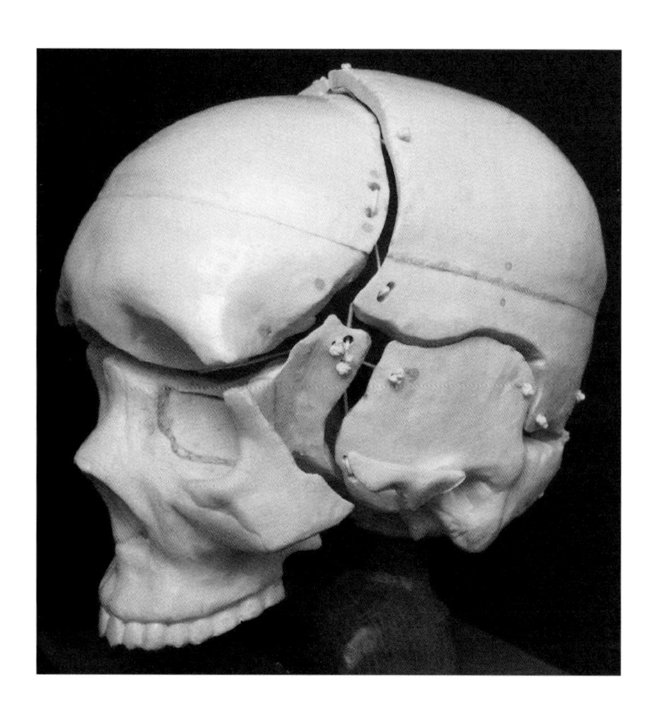

Figure 8.1
A tensegrity model of the cranial vault with the bones as struts, and elastic cord representing the internal dural membrane (dura mater).

the facial bones sitting in front (Figure 8.1). Until recently, the general opinion was that the growing brain must be pushing the bones outwards during development, but it is now known that this is not the case (Mao, 2005). In addition, some practitioners described as cranial osteopaths and craniosacral therapists, etc. have long recognised the significance of a small amount of flexibility between the bones (Sutherland, 1939; Parsons & Marcer, 2005) and used their findings in the diagnosis and treatment of a wide variety of conditions. However, the underlying details that could explain this have generally remained poorly understood.

Although the human skull is generally considered to be a solid box it is actually made up of twenty-two bones, with eight of these contributing to the cranial vault, most of which remain quite distinct throughout life (Sabini & Elkowitz, 2006). The spaces between the bones (sutures) are then filled with fibrous tissue and a tough dural membrane (dura mater) lines the internal surfaces of the vault and cranial base.

A model of the cranium as a tensegrity structure helps to explain how the bones can enlarge *without* the brain pushing them out, and how the skull can retain a certain amount of flexibility (Cummings, 1994; Scarr, 2008). Before getting into the details of this particular model, however, it is worth taking a brief look at some of the geometry that underlies it, which is essentially based around a sphere (Fuller, 1975, 706.10).

The geometric model

As we saw in Chapter 2, a true sphere is only a mathematical construct and cannot exist in the real world, but the shape that comes closest to matching it is the icosahedron (Figure 8.2a), Fuller's geodesic dome (Figure 1.8, page 8). The omnidirectional symmetries of this shape can also be modelled with a tensegrity-icosahedron (Figure 8.2b). If we replace the straight struts with curved ones (Figure 8.2c) and expand their width so that they become curved plates, they then serve to model the bones of the cranial vault. Although the vault actually consists of more bones than this, the principle remains the same (the facial bones, sphenoid and ethmoid have also been modelled as a single unit out of convenience).

Straight into curves

Of course, the more astute reader will recall that geodesics connect points over the shortest distance in *straight* lines (Figure 2.1, page 13) and might be wondering from where these curved struts have suddenly appeared. If we

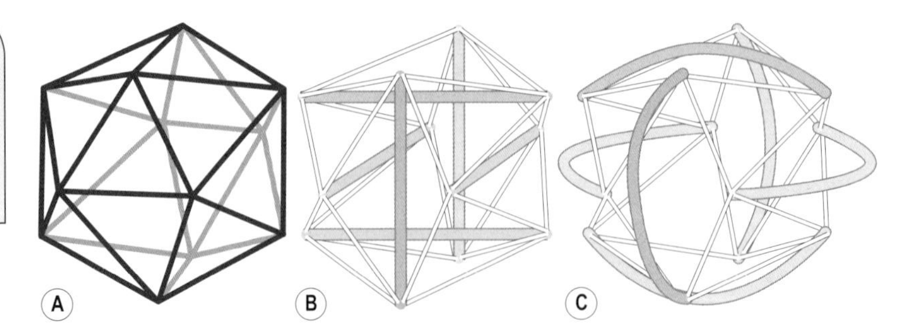

Figure 8.2
A Icosahedron; B T-icosahedron; C T-icosahedron with curved struts (re-drawn from Scarr, 2008).

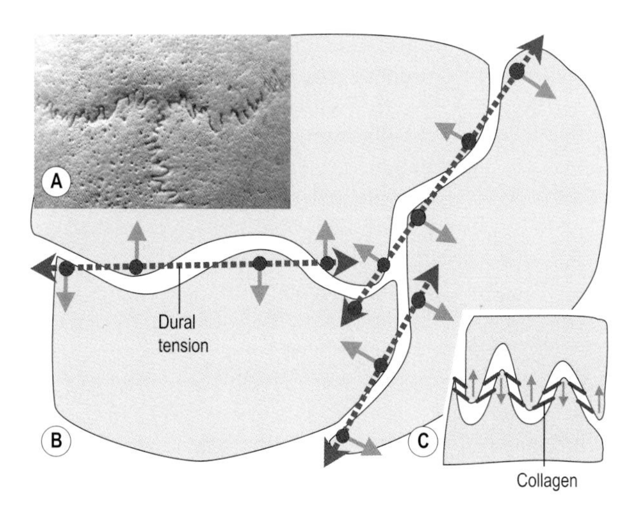

Figure 8.3

A A view of the junction between the sagittal and coronal sutures on the top of the head (bregma) showing the large curves and smaller serrated interdigitations; B schematic diagram of the internal bony surfaces, showing the tensioned dural membrane (dotted arrows) attached around the large sutural convexities, and that has the effect of pushing the bones apart (grey arrows); C a similar effect occurs with collagen in the serrated sutures. (Re-drawn from Scarr, 2008).

invoke the high-frequency icosahedron (Figure 1.7, page 7) as an even closer approximation of a sphere, and note (for now) that straight lines are just a special case in geodesic geometry (Fuller, 1975, 702.01), the appearance of curves at higher levels in biological hierarchies will start to make a bit more sense. To find out more we will have to wait until Chapter 10!

Anatomical basics

Returning to the biology, it should be explained that the underlying dural membrane in infants is attached more firmly around the *margins* of the bone; and in this model the tensioned cables (elastic cord) are also fixed around the bone edges, in particular, to the convexities on either side of the sutures (Figure 8.3b). As the tension vector in the membrane pulls on these convexities it then has the effect of causing the bones to be *pushed* apart, perpendicular to the tension line, and the alignment of collagen in the smaller serrations has a similar effect (Jasinoski et al., 2010) (Figure 8.3c). Incidentally, these large and small sutural curves have a fractal relationship between them (Yu et al., 2003), i.e. they are self-similar at different size scales and geometrically related to each other (Figure 3.9, page 30). To fully appreciate how this model compares to the real thing, we should look at some of the underlying physiology.

Embryonic development

During embryonic development, the growing brain pushes on the surrounding ectomeninx membrane and, after about eight weeks, the bones condense and grow within the membrane (Figure 8.4a). In doing so, they separate it into an outer periosteum and inner dura mater (Figure 8.4b), so that by the time of full-term birth the bones are in close apposition and separated by the sutures (Pritchard et al., 1956) (Figure 8.5).

Bone growth is regulated by cyclic interactions between osteoblasts, which are bone-making cells that cluster around the growing bone edges, and dendritic cells in the dural membrane beneath the sutures (Figure 8.5). These osteoblasts produce chemicals (mitogens) that diffuse towards the dural cells, and together with changes in tension in the dural membrane cause their tensegrity cytoskeletons to stimulate changes in cell function, similar to the description in figure 5.5e (page 47). The dendritic cells then transform into motile cells that migrate towards the bone edges and mature into osteoblasts that increase bone growth. This bone growth cycle then continues up until about seven years of age, with fibrous tissue in the sutures preventing the majority of the bones from fusing together.

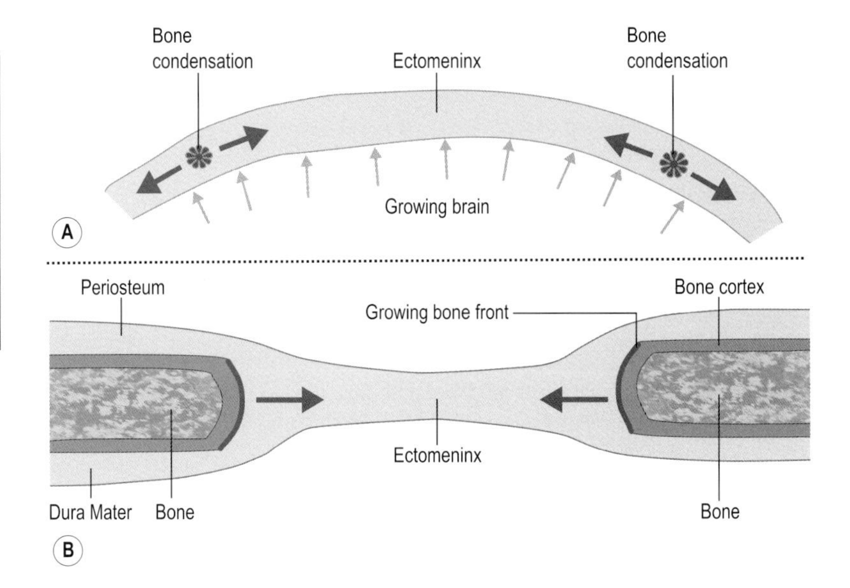

Figure 8.4
A The growing brain initially pushes on the ectomeninx membrane (arrows) in which bones condense; B the bones grow and separate the ectomeninx into an outer periosteum and inner dura mater membrane.

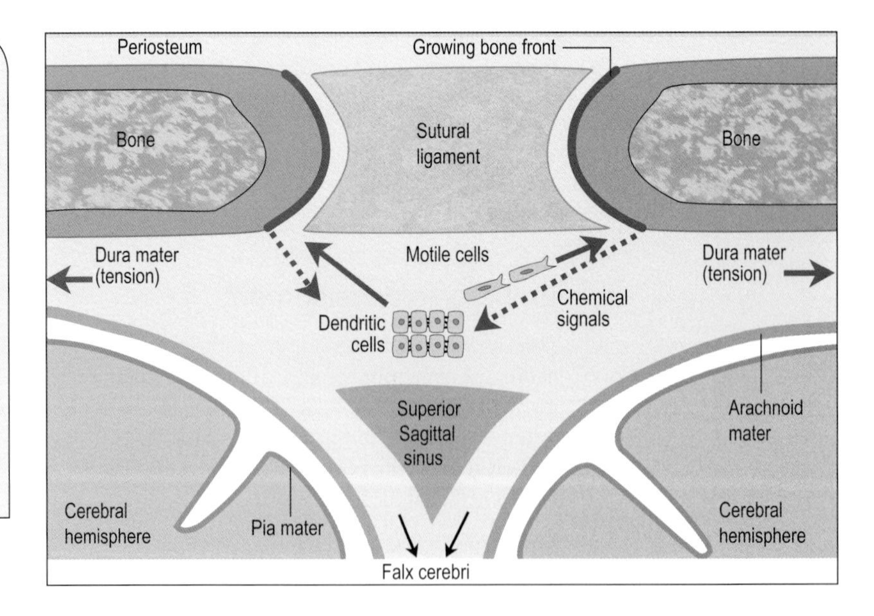

Figure 8.5
Schematic diagram of the sagittal suture showing the cyclic relationship between chemicals produced by osteoblasts at the bone fronts (dotted arrows), dural tension and dural cells; new motile cells then migrate to the bone fronts and form new osteoblasts that increase bone growth (coronal section, not to scale) (re-drawn from Scarr, 2008).

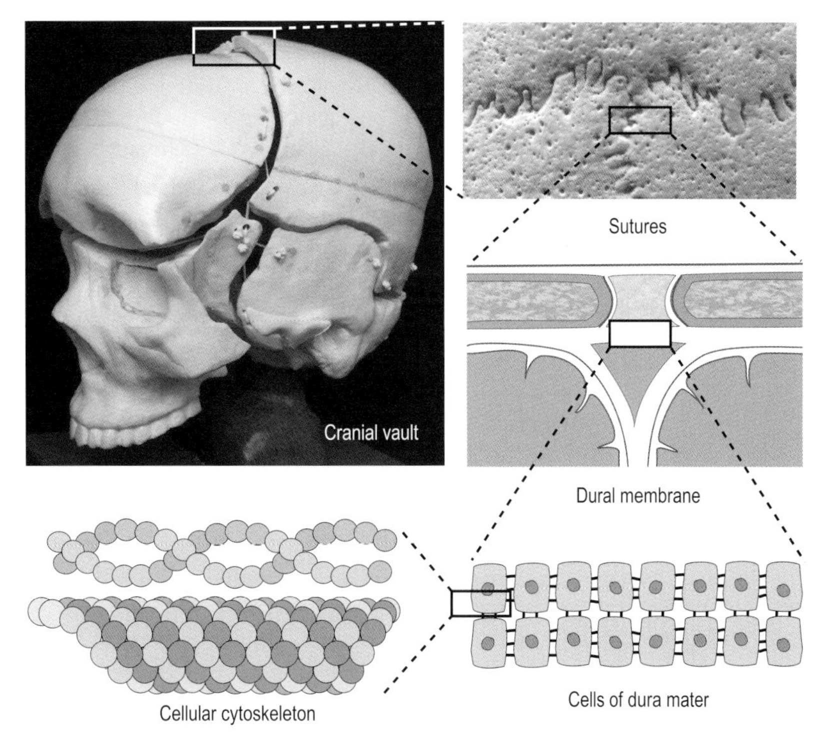

Figure 8.6
A biotensegrity hierarchy functionally couples helical molecules in the cytoskeleton to cells in the dural membrane, osteoblasts that regulate bone growth and stability of the entire cranial vault.

The integrated cranium

The main significance of tensegrity to the cranial vault is that this particular configuration would allow the bones to enlarge and remain stable *after* the initial eight week period, i.e. they could 'float' around the brain rather than being physically pushed out by it (although it is recognised that the brain must have some influence). It would also enable the skull to retain some flexibility and respond to the mechanical demands of external muscular and fascial attachments. The bones then form a dome, with a hierarchical micro-structure (that provides protection to the brain and is comparable with the stability of Fuller's geodesic dome) (Figure 1.8, page 8) and flexible compression struts that accommodate growth; but it is the intrinsically tensioned membrane that integrates and regulates the entire structure from the cytoskeleton upwards (Figure 8.6). The biotensegrity model thus integrates cranial anatomy and physiology at multiple hierarchical levels and helps to explain some of the mechanics of bone motion and associated pathologies (Parsons & Marcer, 2005, p.201; Scarr, 2008).

Cranial pathologies

The bones of the vault are further stabilised by vertical and near-horizontal membranous sheets (falx and tentorium), which separate the cerebral hemispheres and cerebellum and attach to the base of the skull, and are continuations of the dura mater lining the inner surfaces of the bones (Figure 8.5). It is then quite feasible that abnormalities in the developing cranial base (sphenoid/basi-occiput) could alter the normal tension pattern in these dural sheets and distort the tensegrity configuration (and vice versa), thus causing malformations in head shape (plagiocephaly). Abnormal dural tension might also cause changes in sutural osteoblast/epithelial interactions that mis-regulate bone growth and lead to premature fusion (craniosynostosis) that necessitates surgery. Further studies from a biotensegrity perspective may help to resolve these issues and contribute to future treatments.

The value of the tensegrity-icosahedron in describing biological structure at every hierarchical level has been recognised from the very of beginning of the biotensegrity concept, and has become the archetypal model (Ingber, 2008; Chen et al., 2010; Levin & Martin, 2012); and we have also seen that the bicycle wheel follows the same principles and is particularly useful in describing the lungs of birds.

The avian lung

The respiratory system of the bird is an exceptionally efficient gas exchanger that processes the very large amounts of oxygen required to sustain flight, but it differs substantially from the human respiratory system. Although the volume of the bird lung is about 27% less than that of a mammal of similar body mass, the respiratory surface area is about 15% greater. The reasons for this efficiency are considered to be its particular geodesic design and hierarchical tensegrity configuration that mechanically couples each part into a functionally unified structure (Maina, 2007).

Hierarchical anatomy

Unlike mammals, the avian lung is attached to a rigid ribcage, which means that its volume changes little during the respiratory cycle (1.4%). Instead, separate air sacs act like bellows that cause unidirectional and continuous ventilation. The air passages are arranged in a hierarchy, with two-thirds of the lung volume taken up by several hundred parabronchi that are connected to primary bronchi (Figure 8.7), and each parabronchial 'tube' has a central lumen with numerous polygonal openings (atria). These then give rise to several funnel shaped ducts

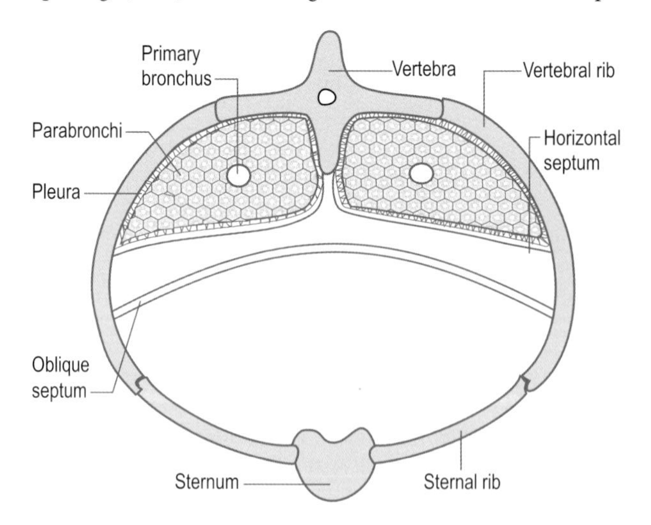

Figure 8.7
A cross-section through the thorax of a bird (re-drawn from Maina, 2010).

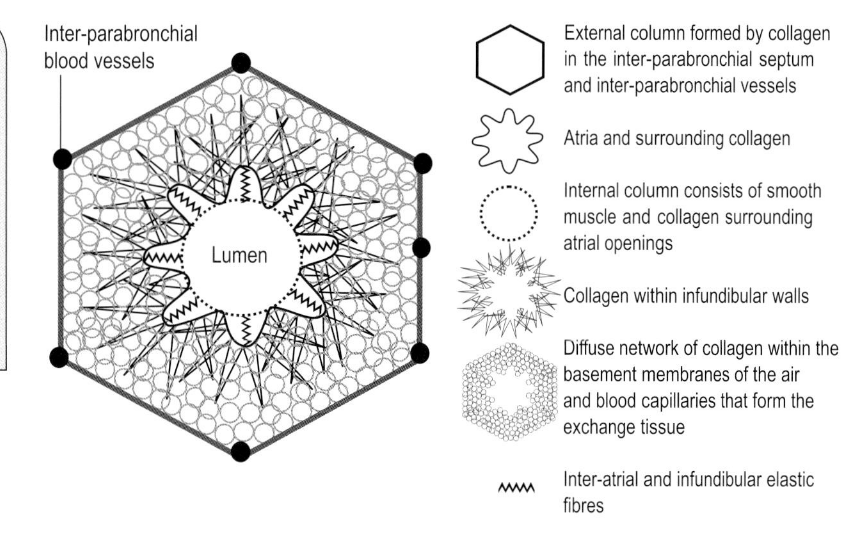

Figure 8.8
Schematic diagram showing a cross-section of a single parabronchus, with the hexagonal outline and exchange tissue suspended between the external and internal columns (re-drawn from Maina, 2010).

(infundibulae) that terminate in multiple air capillaries, the terminal respiratory units (Figure 8.8). Both blood and air capillaries anastomose and interdigitate to form a tightly packaged and complex three-dimensional network.

Each parabronchus develops from a single endothelial cell, with each one assuming a hexagonal shape due to the pressure from surrounding cells, and a central lumen that enlarges during development (Figure 8.9); and whilst this packing arrangement makes the most economic utilization of space and materials (Figure 2.4, page 14) it also maximises the respiratory surface area.

During development, intertwined bundles of smooth muscle, collagenous tissue and elastin form an intricate helical arrangement that surrounds the atria where they open into the central lumen, with collagen fibres continuing into the inter-atrial and infundibula septa as part of the basement membrane surrounding cells of the exchange tissues. The fibres surrounding the atrial openings then form an *internal* parabronchial column (that reinforces the lumen) and continue through the infundibula septa, exchange tissues and inter-parabronchial septa to form an *external* parabronchial column (Figure 8.8).

The tensegrity wheel model

The parabronchi are thus in a dynamically tensioned state, with the inward pull of the atrial muscles (internal column/wheel hub) ultimately balanced by the inter-parabronchial septa (external column/wheel rim) and surrounding parabronchi. The exchange tissues and associated septa are suspended between these two columns like the spokes in the tensegrity bicycle wheel (although the inter-parabronchial septa are not definable in some species), and form an integrated unit that prevents the air capillaries from collapsing under compression and blood capillaries from distending with over-perfusion.

This hierarchical structural array extends down through the muscles and collagen fibres surrounding the parabronchi to the atria and infundibular septa,

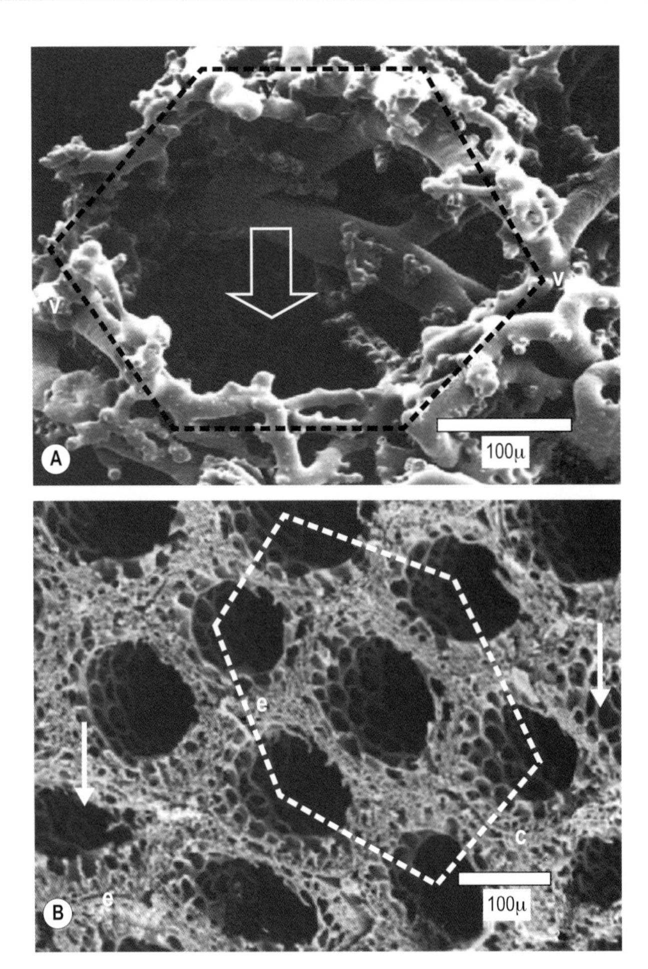

Figure 8.9
Parabronchi of domestic fowl, *Gallus gallus* var. *domesticus*, showing: A a cast of inter-parabronchial blood vessels surrounding the central lumen, arrow indicates the lumen; B the parabronchial packing arrangement showing atrial openings and exchange tissue; scale bars 100 μm. Reproduced courtesy of © Maina, 2010, Elsevier.

and transmits and dissipates mechanical stresses. The tensegrity model explains how small and delicate units like the air and blood capillaries can maintain their mechanical integrity and, along with other refinements, has allowed the avian respiratory system to develop into a finitely compliant gas exchanger (the lung) and highly compliant ventilator (the air sacs) (the same tensegrity principle has also been described in the mammalian lung (Weibel, 2009).

The story so far

Throughout this book we have considered geometry, anatomy and biomechanics at multiple size scales using simple tensegrity models that are essentially sticks and bits of string. Defining precisely which anatomical structure is a cable and which a strut in a biological context can sometimes seem quite complicated, because the 'cables' and 'struts' don't *always* appear as nice straight lines and the terminology does not always fit very well. However, if we accept that models will always be simplifications of the real thing yet still have value in demonstrating the *principles* of tensegrity in biology, we can start to examine the biotensegrity model in a bit more detail.

9 A closer look

The 'design' of plants and animals... did not just happen. As a rule, both the shape and materials of any structure which has evolved over a long period of time in a competitive world represent an optimisation with regard to the loads which it has to carry and to the financial or metabolic cost. J. E. Gordon (1978, p. 303)

Biotensegrity provides a window into understanding the structure and mechanics of complex biology at every hierarchical level, and its essence is structural and functional interdependency between components at *multiple* size scales (Figure 9.1). The stick-and-string models are then just simplifications

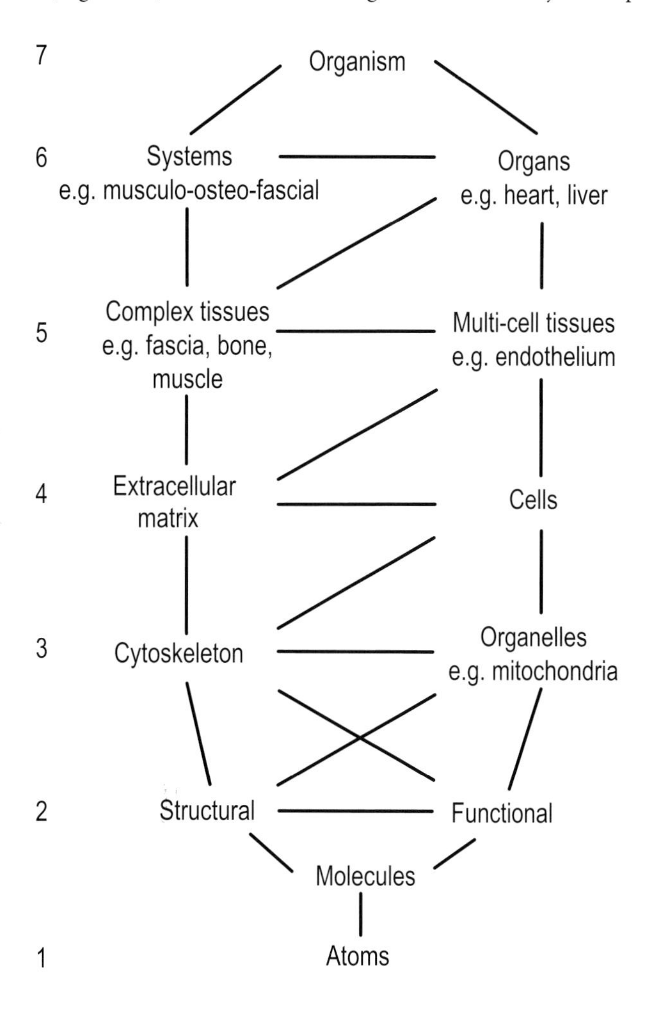

Figure 9.1
An example of an integrated structural and functional hierarchy

of complex biological hierarchies that have been stripped down to a bare minimum, and terms such as 'tension and compression' and 'cables and struts' are used to indicate the structural forces within them. However, while tensegrity models all follow the same basic principles, and the previous chapters have highlighted some anatomical consequences of these, they can sometimes seem rather limiting to the subtleties and complexities of biology if applied too rigidly.

Unlocking the terminology will make things a little clearer – it is time to pull things apart!

Tension and compression

In a tensegrity chain, every cable is under tension and every strut under compression, but the whole structure can be stretched to form a cable or compressed to form a strut at the next level in an even larger hierarchy (Figures 3.6 and 3.7, pages 29 and 30). In a biological context, each small cable and strut would also have its own distinct material properties that affect how it responds to tension and compression, and the behaviour of any particular part will then depend on the huge number and variety of smaller components that it is made from. At the smallest size scale, however, other terms are more useful.

Attraction and repulsion

Both Snelson and Fuller described tensegrity as *continuous tension and discontinuous compression*, with a clear distinction between forces, but these terms are usually applied to structures that are relatively large. At the smallest of scales, atoms interact to form molecules that are tensegrities in their own right, but here the 'cables' and 'struts' are invisible (Zanotti & Guerra, 2003; Edwards et al., 2012). These atomic configurations assemble spontaneously through Van der Waal forces that both *attract* and *repel* depending on the relative position of different atoms or a change in pH; covalent bonds that attract but maintain a fixed distance apart; hydrogen bonds that attract but are flexible in spacing; and steric repulsion, where atoms are kept a distinct distance apart; etc. (Sadoc & Rivier, 2000). As tensegrity structures are 'the physical representations of the invisible forces within them', it is actually more meaningful to use the terms *attraction* and *repulsion* when considering molecules at the nano-scale; and although this might be considered as mere semantics, different words can imply nuances in meaning that are more useful in certain biological contexts.

Pull and push

Terms such as *pull* and *push* imply active forces, and are sometimes used in computerised modelling, but they are not necessarily synonymous with tension and compression. The compliant properties of biological structures vary considerably and interact in a huge number of different ways, and are far more complex than those of our simple stick-and-string models. In these structures, tensioned cables *pull* on the nodes and compress the struts (Figure 9.2a), but there might be situations when the struts can actively *push* the nodes apart and increase cable tension (Figure 9.2b). Because biology is not restricted to using

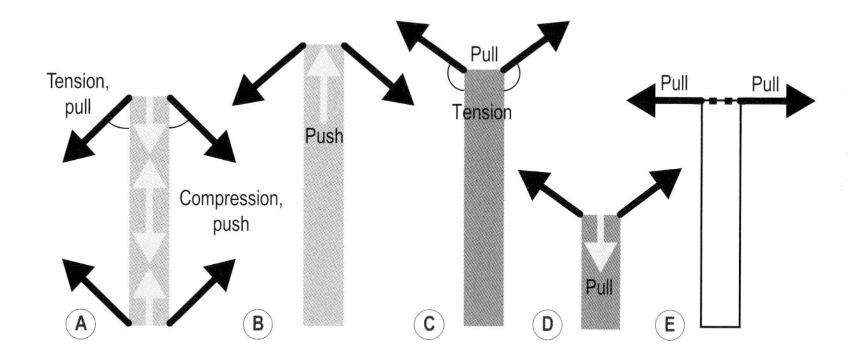

Fig. 9.2
A a strut is normally compressed by the tensioned cables *pulling* on the nodes at either end; B a strut exerting a *push* on the nodes and increasing cable tension and length; C cables *pull* the nodes apart and tension the strut; D a contracting strut *pulls* the nodes together and increases cable tension; E cables that are perpendicular to the strut do not compress it, although they would if attached on opposite sides (dotted).

any particular number or arrangements of cables, some of these could also *pull* the nodes apart and make the strut longer – as part of a functional mechanism that regulates length (Figure 9.2c) – or the struts might shorten and become tensioned as the cables attempt to *pull* their ends apart (Figure 9.2d).

Microtubules, that are generally thought of as being compressed, can increase in length in response to a *pull* from the ECM (Figure 9.2c) (via integrins that cross the cell membrane) (Figure 5.3, page 45), while actomyosin motors can modulate this by increasing actin tension that *pulls* on the ends of the microtubules and shortens them (Figure 9.2d), thus maintaining a dynamic balance (Ingber, 2003a). Cellular growth can *push* against surrounding cells and tissues, which then become stretched (tensioned) (Figure 9.2b) and switch function (Figure 5.5, page 47), or cells can contract and *pull* on tissues that now become tensioned in a different direction (Figure 9.2d) (Henderson & Carter, 2002; Blechschmidt, 2004, p. 62). The contraction of muscles can also cause a simultaneous increase in width that *pushes* outwards and tensions surrounding fascia (Figure 9.2b), which means that a muscle is both an axially tensioned cable and a compressed strut perpendicular to this.

The possibilities are endless

Essentially, tension is always trying to reduce itself to a minimum and form a straight line, which is one of the most important factors in the self-balancing ability of tensegrities. In a complex biological tissue, all the variations in Figure 9.2 are likely to be operating at different places in different permutations, and it is the balance between all the different *pushes* and *pulls* that automatically determines the final position of equilibrium. This way of looking at things emphasises that biotensegrity is far more than just 'cables-and-struts' but the interaction of forces within a complex spatial network – a dynamic biological structure that has many things happening inside it.

In simple terms, tensioned cables cause the struts to be compressed when they are attached to the nodes at an acute angle (Figure 9.2a), but *pull* the nodes apart when at an obtuse angle (Figure 9.2c); and in a biological context, microtubule polymerisation and the growth of bones, etc. develop in response to the direction of these different vectors. A tension vector (*pull*) thus becomes a compressive one (*push*) when it changes angle, and more tension means more compression. Although cables that are perpendicular to the strut are generally neutral in this respect, this depends on which side of the strut they are attached to (Figure 9.2e).

In the cranial vault model, tension in the dural membrane and sutural collagen (cables) has the resultant effect of *pushing* the bones apart, as the tension tries to *pull* all the bony attachments (nodes) into a straight line (Scarr, 2008; Jasinoski et al., 2010) (Figure 8.3, page 71). Interestingly, most man-made materials that are stretched tend to get thinner in the middle, and those that are compressed bulge outwards (Poissons ratio), whilst biological materials (Gordon, 1978, p. 161) and tensegrity structures frequently show the opposite effect (Figure 3.4, page 28) (Edmondson, 2007, p. 272).

The terms *tension* and *compression, attraction* and *repulsion, pull* and *push* are thus describing the same things, but from different biological perspectives, and recognising their subtle differences will prove useful. Biotensegrity is about the balance of force vectors, and the main role of a 'cable' is to attract and pull things together while that of a 'strut' is to repel, push and hold them apart. Although soft tissues are usually thought of as tensioned cables and hard ones as compressed struts, the difference between a cable and strut is really just relative because they can change from one to another. So, how can we tell the difference?

Cables and struts

The lost strut

Figure 9.3a shows a model made from two T-icosahedra of different sizes joined together, with one of the vertical struts from the smaller model removed and the cables that were attached to its nodes fixed to a vertical cable from the larger one (black), and this now pulls the nodes of the missing strut apart. However, we now have an apparent paradox, because the black line is still a cable with respect to the large model, but *acting* like a strut as far as the smaller one is concerned; and the only difference is that the nodes of this 'lost' strut are now being pulled *apart* by the larger model and *together* by the black part of the cable. We could say that the entire large model has replaced the strut from the smaller one, or the entire small model has replaced the cable from the larger one, but they have actually become united into a new tensegrity configuration.

Remember that we are really describing force vectors, which means that any particular piece of a complex tensegrity could be functioning as both a cable and strut at the same time (depending on how it is considered within the hierarchy), or even change emphasis from one to the other during movement. Although each part might be influenced by multiple tension and compressional forces, one or the other state will always predominate at any particular point in time. The differences between what we call cables and struts in a *biotensegrity* context are thus relative, a useful way of functionally distinguishing between components, but they have their limitations.

Simple evolving into complex

Biological structures must remain stable at each instant of their existence, and the transition from one state to another must have an energetic advantage if

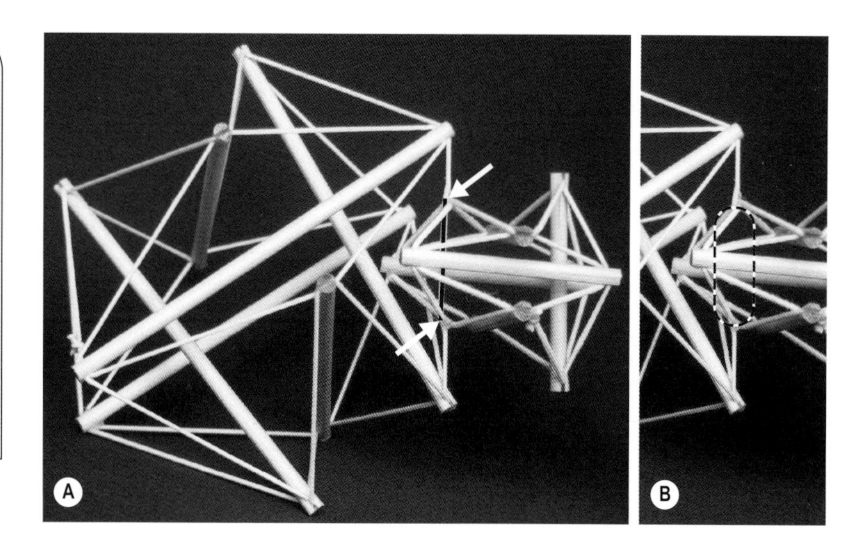

Figure 9.3
A Two tensegrity models (T-icosahedra) of different sizes joined together so that a strut from the smaller one is replaced by part of a cable from the larger (black line between arrows); B removal of this now redundant strut leaves a space (dotted) and a structure with a different position of equilibrium.

their development is to be viable (Levin, 1997, 2006). As this common cable/strut is now structurally superfluous, its removal would cause the rest of the structure to rebalance in a new position of stable equilibrium and make the overall system *more* efficient, in material terms, but with different mechanical properties (Figure 9.3b). It would also leave a space that we might consider equivalent to that between bones in the elbow (Figure 7.11, page 67), or sutures in the cranial vault (Figure 8.3, page 71), with the soft tissues holding the bones *apart*.

Tensegrity structures are thus modular and can be disrupted and enlarged without necessarily compromising their overall integrity – a useful facility during growth and development because more complex configurations can then be energetically more efficient than simple ones. Biotensegrity explains how the structural mechanics of a molecule held together by the invisible forces of attraction and repulsion are not really that different from an animal that contains hard bones a million times larger. As the same principles apply at every hierarchical level, soft tissues must also have their 'hard' struts.

A search for the missing compression

Fascia is a 'soft' connective tissue that plays an important role in the transfer of tension and, until relatively recently, was generally considered to be little more than mere packing (Van der Wal, 2009). Now the fascia is more widely recognised as a tissue that is continuous with the extra-cellular matrix that surrounds virtually every cell in the body (Ingber, 2006); a hierarchical structure that links every bone, muscle, nerve, blood vessel and skin, etc. into a single functioning unit (Schleip et al., 2012; Guimberteau, 2005) and part of a global biotensegrity system (Levin & Martin, 2012). Tensioned fascia implies tensegrity cables, but once again the distinction between cables and struts becomes blurred; where is the compression?

To answer this, we should consider a sheet of fascia tensioned between two bones (Figure 9.4), where two points aligned along the direction of stretch (x and y) are being pulled apart. Note also that one of the functions of a strut is in keeping the nodes apart i.e. maintaining a separation (Figure 9.2a), so, although the points x and y are held apart by the tensioned fascial 'cable' on either side, the piece of tissue between them is functioning like the virtual strut in Figure 9.3a. This tensioned 'strut' could then form part of a tensegrity configuration, and act as such to other parts lower down in the fascial hierarchy, as represented by the tensegrity-icosahedron model shown.

The balance of forces between parabronchiolar tissues within the avian lung (Figure 8.8, page 75), and indeed between groups of cells with differing levels of stiffness, is essentially similar to this fascial example. It is also significant that the gleno-humeral capsule (which is really a specialised sheet of fascial tissue) cannot transfer tensional stresses in certain directions if parts of it are damaged or missing, even if those stresses would not be expected to pass through those parts (Moore et al., 2010). This makes perfect sense if the micro-structure of the capsule is a tensegrity configuration.

Fascia, and in fact all soft tissues, can thus be considered as networks of cables and struts at multiple hierarchical levels, but it is suggested that they only function properly because they are integrated into the global tensegrity system, where compression is provided by proteoglycans, microvacuolar 'bubbles' and local changes in the stiffness of the extra-cellular matrix, etc. at the micro-level (Guimberteau, 2005; Kapandji, 2012), and higher level compression is provided by the bones and soft-tissue compartments, etc.

It is all about perspective

Labelling any biological tissue as a 'cable' or 'strut' is clearly just a matter of perspective, because each one consists of a complex network of energy vectors at multiple hierarchical levels (Figure 9.1), and it is the interactive balance between them that determines how the structure behaves mechanically. Muscles actively

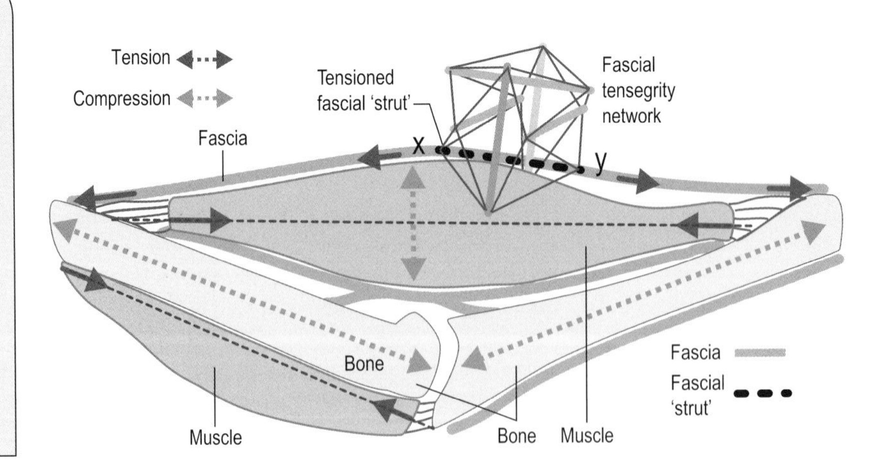

Figure 9.4
A schematic diagram of a 'typical' joint showing a fascial sheet stretched between two bones, and some of the tension and compressional forces. The fascia between the points x and y (dotted) is held in tension but behaving like a tensegrity strut to other parts further down in the tissue hierarchy (tensegrity-icosahedron).

generate tension when they contract and create variable length compression struts across their width (Figure 9.4), and this would then be likely to change the tension in associated fascia (Huijing & Baan, 2003). In fact, a strut should be considered as *any* structure that maintains a separation, such as across bones, muscles, blood vessels, fascial compartments and even single cells.

While these forces of tension and compression always act in straight lines – it is easy to think of cables and struts in this respect – curves are also a common feature in biology; and it is now useful to consider where they come from in a hierarchical context.

Straight or curved

One of the principles of geodesic geometry is that any two points are connected over the shortest path, and in our familiar three-dimensional Euclidean space (E^3) this is always a straight line. However, a straight line is just a special case in geodesic geometry and is just one of three geometries that describe space (Fuller, 1975, 702.01), the others being *Spherical* (S) and *Hyperbolic* (H), both of which are curved (Terrones et al., 2001; Bowick & Giomi, 2009). Curves can be a means of minimising energy, but generate damaging internal stresses if they are subjected to excessive force from outside (Figure 9.5) and must be stabilised in ways that limit the disastrous consequence of this.

A curved structure that is tensioned creates internal shear stresses that cause it to buckle and fracture, if the load gets above a certain limit (Figure 9.5a–c), while a straight one that is under too much compression can bend into a curve with the same result (Figure 9.5d–f). The external forces produce bending moments, just like a lever would, and these then generate internal force vectors (shear stresses) that slide across each other in different directions. Depending on how stiff the structure is, there comes a point where these stresses become so great that the material becomes fatigued and collapses.

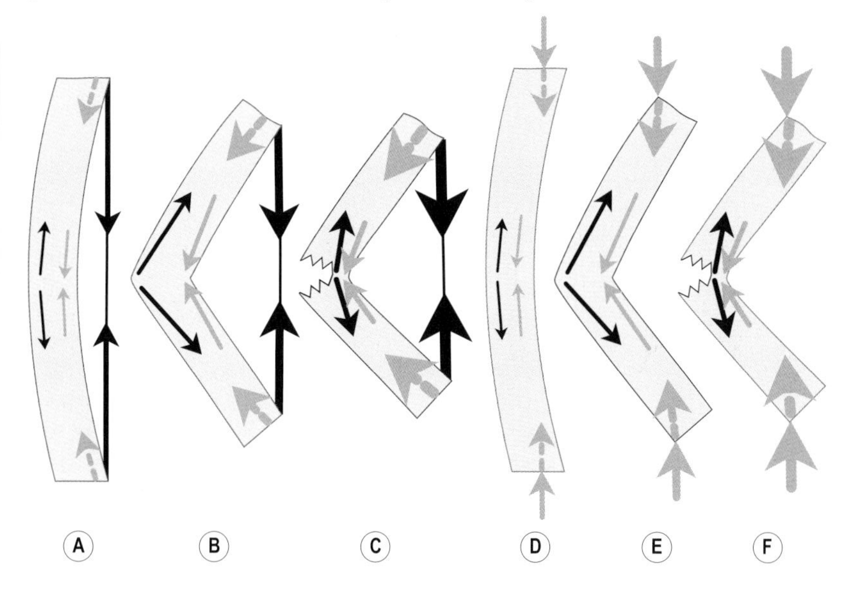

Figure 9.5
Schematic diagram showing how *externally* applied tension A–C (large dark arrows) and compression D–F (large light arrows) can generate *internal* shear stresses in curved structures.

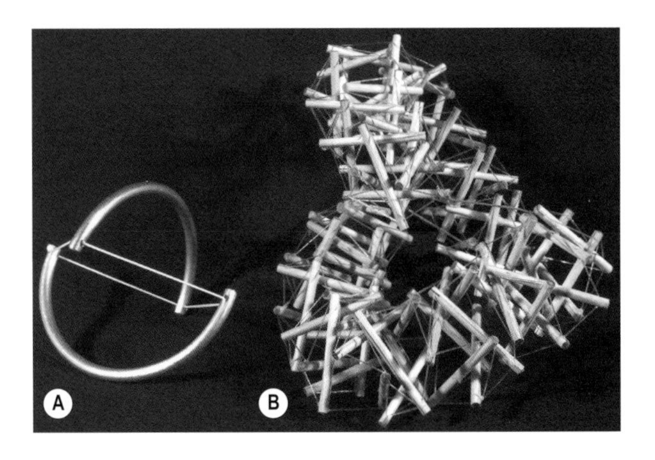

Figure 9.6
A A tensegrity model with rigid curved struts (90° to each other) would buckle and fracture if the tension increased above a certain limit (Figure 9.5a–c); B but a similar structure with the curved struts modelled as a tensegrity hierarchy has no internal shear stresses and easily distributes the load. © Rory James

Reducing the stress

Biology, however, reduces the consequences of shear stresses by utilising the features of hierarchies, where damaging stresses are transferred down to a lower level and eliminated in the balanced attraction and repulsion between atoms (Levin, 1982, 2002). Tensegrity shows how a structure can distribute stresses throughout the system and remain resilient to the effects of external forces (Figure 9.6), where even the connections between the cables and struts (nodes) are hierarchies in their own right (Fuller, 1975, 740.20) (Figure 3.7a, page 30).

Compressed microtubules might then appear to bend and buckle but remain stable because they are structurally balanced by tensioned intermediate filaments (Figure 5.3, page 45) (Brangwynne et al., 2006). Potential bending stresses in long bones can be reduced by the balanced tension of muscles surrounding them (Sverdlova & Witzel, 2010) and by the resiliency of their internal hierarchy, where a coupled deformation mechanism between collagen and hydroxyapatite effectively redistributes the strain energy and protects against micro-fractures (Gupta et al., 2006).

Ultimately, biological curves are not continuous but consist of lots of force vectors acting in straight lines within a complex hierarchy, and these give the *appearance* of curves at a higher scale (Figure 9.7). To appreciate where this leads to, we need to take another look at that basic geometry.

Spherical geometry

In the world of geometry, the vertices of all the Platonic polyhedra are equidistant from their centres (Figures 2.5 and 2.6, page 15), just like the surface of a sphere, the most fundamental of all shapes. The icosahedron, as one of these, encloses the largest volume within the smallest surface area of any structure apart from a sphere; and one with a high enough frequency becomes virtually indistinguishable from a sphere (Figure 9.8). Natural spheres are then best considered as high-frequency icosahedra made from multiple distinct

Figure 9.7
A curved chain of tensegrity-icosahedra consists of many straight lines at a lower hierarchical level. © Rory James

components (Fuller, 1975, 515.00; Edmondson, 2007, p. 265), and the shortest distance between any two points now *looks* like a curve.

A true sphere is really a two-dimensional curved surface (S^2) embedded in a 3-D space (Terrones et al., 2001), but it is only from convenience and a mathematical perspective that we describe these geodesic lines as curved, because *all curves in biology must be straight lines at some smaller scale.* Levin emphasised that the icosahedron is the minimal-energy shape that compliant biological structures will always try to reduce themselves into, even though they may be prevented from fully achieving this because of other constraints, and that the tensegrity-icosahedron should be considered as the finite-element and archetypal model of biological structure (Levin & Martin, 2012). Spherical geometry then describes energy-minimising surfaces such as soap bubbles (Isenberg, 1992, p. 12) (Figure 2.3b, page 14), micelles (Butcher & Lamb, 1984; Fischer et al., 2011), cell membranes (Zhu et al., 2007) (Figure 5.3, page 45) and spherical viruses (Twarock, 2006), etc. but their curved appearance is structurally rationalised because they are all made from many small parts connected in straight lines at a smaller scale (Figure 9.7).

The nuances of anatomy

While such terms as 'tension and compression' and 'cables and struts' are useful in describing tensegrity models, such simplicity clearly hides many details and nuances that are actually quite significant from a biological perspective. Thus, we have seen that 'soft tissues' contain both cables *and* struts (Figure 9.4), and how slight differences in their mechanical properties will determine how forces

Figure 9.8
The sub-division of each icosahedral face into smaller triangles creates higher frequency icosahedra with the *appearance* of a sphere.

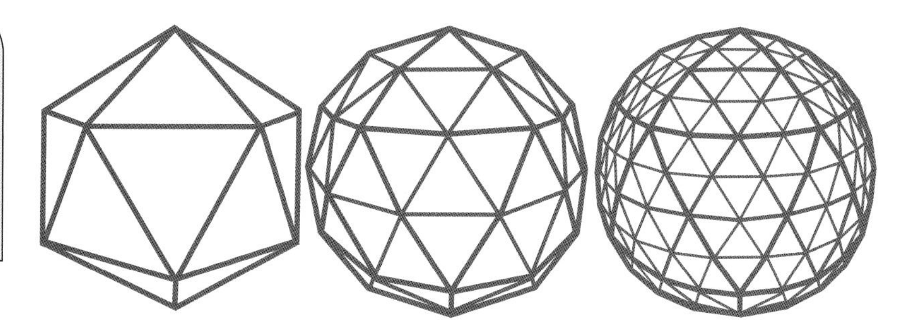

are transferred between them. Biological 'struts' might also *appear* to compress each other at one particular hierarchical level but, when observed in detail, are clearly separate. We have also seen that curved structures that might *appear* to be riddled with shear stresses remain perfectly intact because they dissipate them down through the hierarchy. Most importantly, anatomical shapes develop as they do in response to the complex interactions between the forces within them, and our preconceptions about bones, muscles and fascial sheets, etc. must be modified in the light of this. Curved geometries are then particularly useful because they show how complex patterns and shapes can develop within a different spatial framework, and as a more energy-efficient alternative to simple ones (Fuller, 1975, 706.10).

The following chapter looks at more of those natural shapes, with theoretical geometry providing a rich multi-dimensional canvas upon which complex biological shapes and biotensegrity can develop.

10 'Complex' patterns in biology

The harmony of the world is made manifest in Form and Number, and the heart and soul and all the poetry of Natural Philosophy are embodied in the concept of mathematical beauty. D'Arcy W. Thompson (1961)

Nature always does things in the simplest way possible, and atoms interact with each other to form simple geometric shapes because they are the most energy-efficient arrangements. Similar shapes then appear at multiple size scales because they are based on the same rules. However, trying to follow the increasing complexity within them soon becomes impossible, which is why an understanding of basic geometry and tensegrity is so important.

In Chapter 2, we saw how simple shapes form through the related principles of geodesic geometry, close-packing and minimal-energy (Van Workum & Douglas, 2006). Later chapters showed how the tetrahelix (Figure 6.1, page 49), icosahedron (Figure 3.3, page 27) and their tensegrity counterparts are useful for examining the structure and function of more complex biological shapes, including humans. In addition, the cuboctahedron emerged from the isotropic vector matrix (IVM) (Figure 2.14, page 20) – the triangulated lattice network that fills space completely and includes the tetrahedron, octahedron and cube, with each one appearing within different frequencies of all the others – and, as the vector equilibrium, modelled transformations in shape and energy through the jitterbug (Figures 2.18 and 2.19, pages 22 and 23).

We can now explore further *how* and *why* these simple shapes are so widely spread in nature (and are more than just interesting curiosities), which directly implicates biotensegrity as it is based on the same principles (Levin, 1981). This chapter is *not* about the classroom geometry of abstract proofs, theorems and heavy mathematics, but a deeper view of the basic rules that underlie anatomy and biomechanics. While some of what follows may be unfamiliar it will hopefully take the mystery out of natural geometry.

While the introduction of spherical geometry (Figure 9.6, page 84) showed that curvature can be an energy-efficient means to creating more complex forms, some new shapes and completely different ways of thinking about structure and space can now be introduced, because ultimately, human function is based on the simple rules of physics.

The rhombic dodecahedron

All the Platonic shapes are further interesting because each one can be matched with another that is referred to as its *dual*, which is a particular relationship where the number of vertexes of one shape is the same as the number of faces

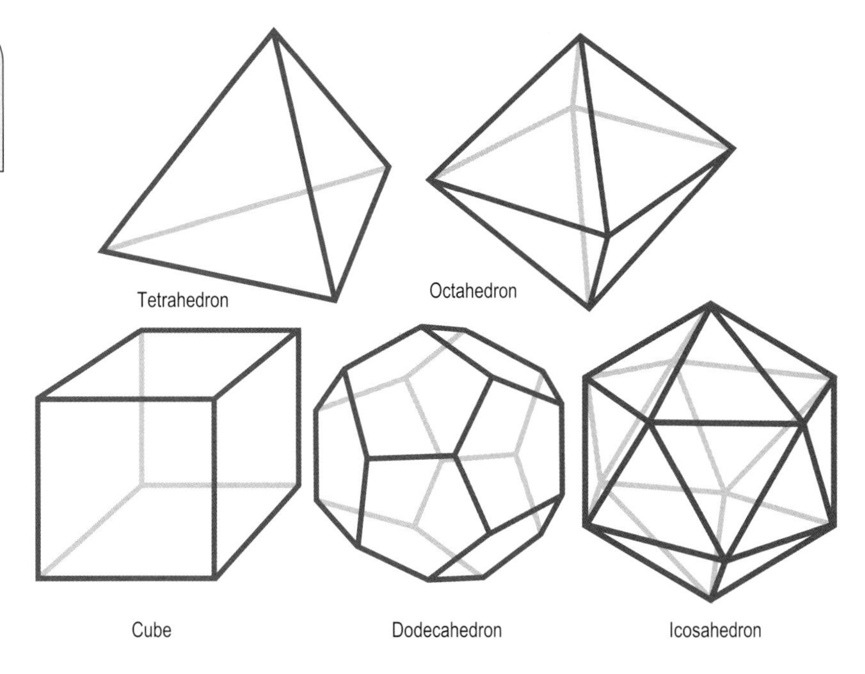

Figure 10.1
The five platonic polyhedra (re-drawn from Scarr 2010).

of the other, and each one can be placed inside the other so that its vertices correspond with the centres of the other's faces. So, the octahedron and cube are duals of each other, as are the icosahedron and *pentagonal* dodecahedron (Figure 10.1). The tetrahedron, as the simplest way of enclosing space, is then the dual of itself (Edmondson, 2007, p. 54) (Figure 2.13, page 19). Although the cuboctahedron is an *irregular* polyhedron because of the mixed triangular and square faces (Figure 2.14, page 20), it also has a dual and this is now a regular shape called the *rhombic* dodecahedron (which naturally occurs in crystals such as garnets), and integrates the different close-packing arrangements of the IVM and icosahedron (Fuller, 1975, 426.10) (Figure 10.2a).

Conventionally, mineral crystals can only have 2-, 3-, 4- and 6-fold symmetries, and thus relate to the IVM, but there is no 5-fold (iron pyrite is a deceptive example) (Figure 2.17a, page 21). However, although the icosahedron and *pentagonal* dodecahedron *do* have this symmetry, they are spatially out of phase with the IVM because of it; i.e. they cannot be arranged in the same close-packing arrangement (Edmondson, 2007, p. 189). The *rhombic* dodecahedron is then particularly interesting because it *does* fill space completely – although in a

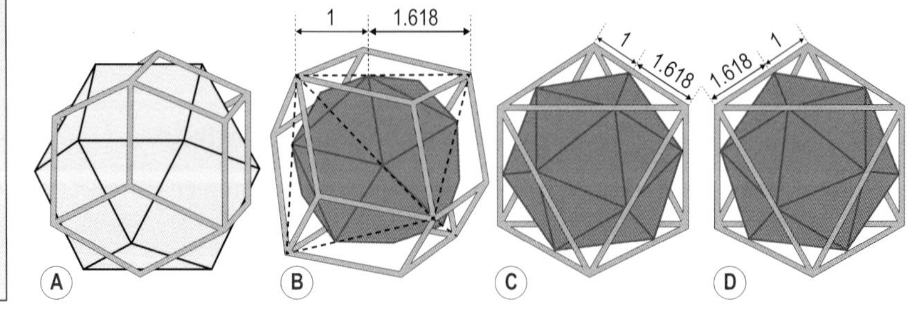

Figure 10.2
A A cuboctahedron enclosed within its dual, the rhombic dodecahedron; B the icosahedron relates to the rhombic dodecahedron through the Golden Mean; and C and D to the octahedron (and thus the IVM) in two different ways (re-drawn from Edmondson, 2007, pp. 61 and 189).

different way to the IVM – and shows how they *all* relate to each other through an even more fundamental relationship: the Fibonacci sequence and Golden Mean (Edmondson, 2007, p. 189).

Figure 10.2b–d shows how the Golden Mean, or Golden Ratio, relates the rhombic dodecahedron with the icosahedron (and its dual) and the octahedron (and all the shapes that emerge from the IVM), and unites both close-packing and space-filling systems (and biological structures are inherently close-packed space fillers). Although this might seem like it is getting a teeny bit complicated, it is the overall principles that are important and it will soon start to make sense.

The Fibonacci sequence and Golden Mean

The Golden Mean is an irrational mathematical constant that describes a continuous balanced proportionality, i.e. the same relationship is maintained between different components no matter where they are in the structure, and it is fundamental to the geometry of shapes with multiple symmetries (Figure 10.3) (Stewart, 1998).

Flower heads display parallel spirals that run in opposite directions as a result of the most efficient packing arrangement of objects in their growing tips (Douady & Couder, 1996), and the number of elements within each opposing spiral is nearly always two consecutive numbers of the Fibonacci sequence, where each new term is the sum of the two preceding ones (0,1,1,2,3,5,8,13,21,34…). The ratio of any two consecutive numbers then approximates to the Golden Mean (0.618), and becomes closer as the sequence gets higher. It also relates to the tetrahelix (Sadoc & Rivier, 2000) and platonic solids, and frequently appears in the proportions of biological structures.

The pattern on the side of a pineapple (Figure 10.3c) and cactus relate to the same sequence (Atela et al., 2002), with the left- and right-handed helixes having different angles (Figure 6.10d, page 54). In animals it is found in the nautilus (Stewart, 1998) (Figure 10.3d), the arrangement of microtubule components (Hameroff & Tuszynski, 2004) (Figure 6.3), the position of coronary artery lesions (Gibson et al., 2003), the packing arrangement of tropocollagen molecules (Charvolin & Sadoc, 2011) (Figure 6.7, page 52) and (maybe) scales on the limbs of the pangolin (Figure 6.13b, page 56) and myofascia of humans (Scarr, 2013). The real significance of the Golden Mean from a biological perspective is that it describes the most efficient way of producing helical and spherical shells from parts that are all the same, and bonded together in the same way (and we have already seen that the human body is full of these shapes), and these are then said to be *equivalent*.

Figure 10.3
A and B Flower heads; C pineapples and D the nautilus display spiral/helical patterns related to the Fibonacci sequence and Golden Mean.

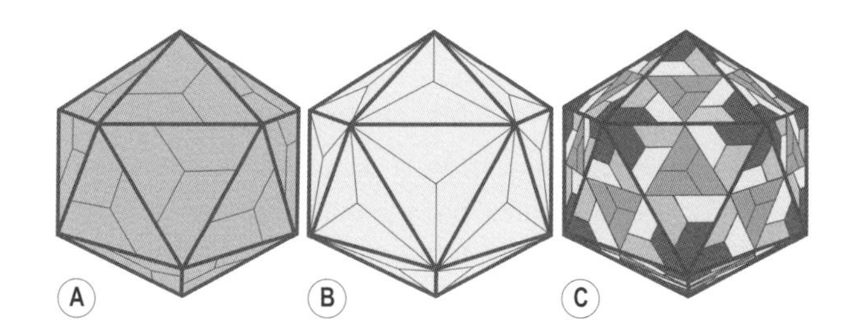

Figure 10.4
A and B the maximum number of proteins that can be connected equivalently in a virus capsid is 60 (3x20); C a virus capsid showing how proteins can be connected in different combinations (quasi-equivalently) around an icosahedral template.

Equivalence

The principle of equivalence implies that components are arranged symmetrically, and it is interesting that the *only* shapes that can accommodate it have surfaces based on cylinders (helixes) and the Platonic shapes (spheres) (Van Workum & Douglas, 2006), but biological shapes are frequently *not* identical and unlikely to match the points on a geometric lattice precisely because of this. Evolution, however, has evaded this inconvenience through the device of *quasi-equivalence*, where different parts contort slightly as they relate to the geometric template and bond together in different ways (Figure 10.4).

Quasi-equivalence and the spherical viruses

The principle of quasi-equivalence was first recognised in the outer capsid shells of the spherical viruses, which we now know are all based on the icosahedron (Caspar, 1980). Although this shape is enlarged by the subdivision of its triangular faces, the maximum number of components that can be joined in an equivalent manner is only sixty (20 faces × 3-fold symmetry). Increasing the number of subunits beyond this limit means that the bonding between them cannot be the same, with some forming pentamers and others hexamers, and they are then *quasi-equivalent* (Figure 10.4c). Quasi-equivalence explains how enzyme complexes such as pyruvate dehydrogenase can have cubic and dodecahedral cores (Izard et al., 1999), and its functional value to virus infectivity has also been described using the tensegrity model (Kovacs et al., 2004; Cretu & Brinzan, 2011).

For many years, the capsid proteins of some spherical viruses *(papovaviridae)* did not seem to fit with the 5-fold symmetry of the icosahedron, but it is now recognised that they do relate to it through the Fibonacci sequence, Golden Mean (Figure 10.5) and a different kind of close-packing arrangement referred to as *Penrose tiling.*

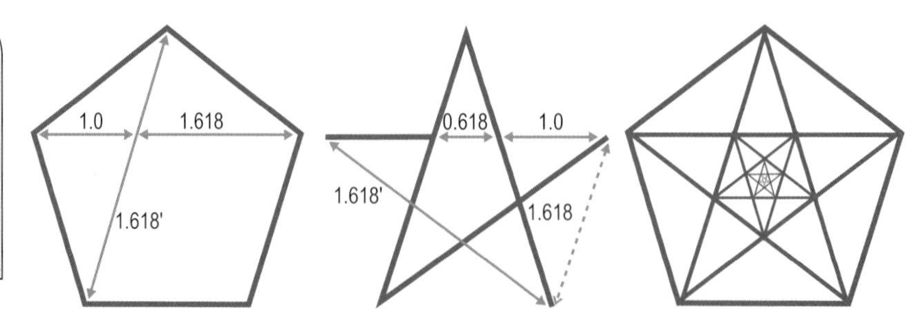

Figure 10.5
The 5-fold symmetry of the icosahedron is intrinsically related to the Golden Mean and repeats itself through a pentagon/pentagram hierarchy.

Penrose tiling

It is well known that the complete filling of a two-dimensional plane with a single regular shape can only be accomplished with triangles, squares, rhomboids and hexagons (3, 4 and 6 sides) (Figure 10.6). Some biological examples of triangulated hexagons were shown in Figures 2.4, page 14 and 8.9, page 76. The resulting pattern is then called 'periodic' because these shapes (tiles) repeat themselves perfectly in any given direction (translation symmetry).

In 1974, Roger Penrose discovered a different tiling pattern that uses two or more different shapes to tessellate a flat plane, but the complete pattern *never* repeats itself in any direction because it is *non-periodic* and lacks translational symmetry (Figure 10.7) (Lord, 1991). Penrose tilings are novel because they have 5-fold symmetry, the same as the icosahedron (Figure 10.5), and use common shapes such as rhomboids, darts and kites, etc; and here is the best bit: the numbers, areas and positions of each shape all relate to each other through the Golden Mean (0.618)!

To ensure that Penrose tiles really are arranged non-periodically, the edges can be matched with specific lines drawn across each tile (Amman bars) and reveal yet another hidden symmetry (Figure 10.8) (Lord, 1991). These bars form five

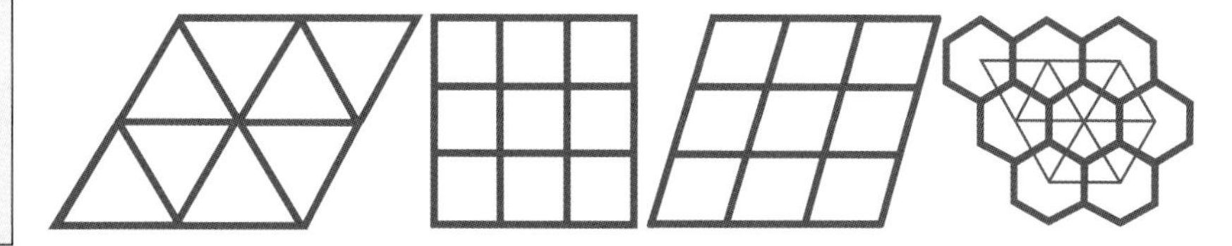

Figure 10.6
The tessellation of a plane with a single regular shape (tile) forms a periodic pattern that repeats itself in any given direction; note the triangulated pattern that underpins the stability of close-packed hexagons (Figure 2.1, page 13).

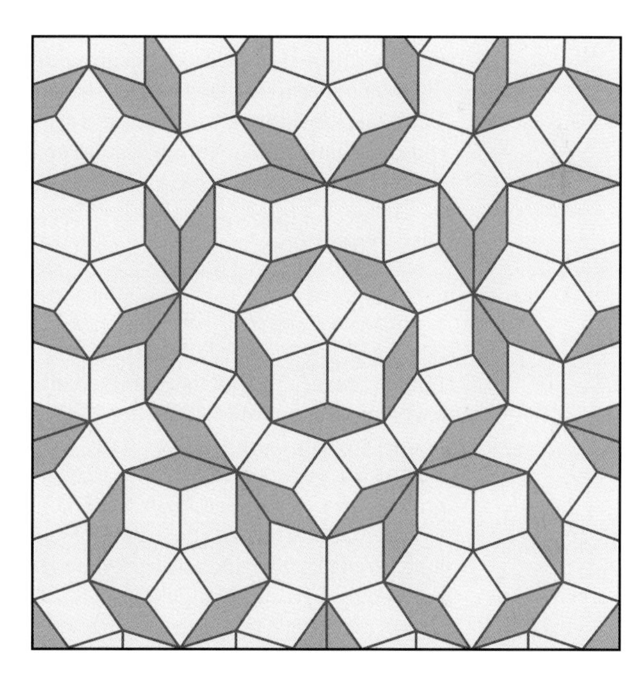

Figure 10.7
A Penrose tiling pattern is non-periodic and *never* repeats itself precisely in any direction (re-drawn from Inductiveload, Wikipedia).

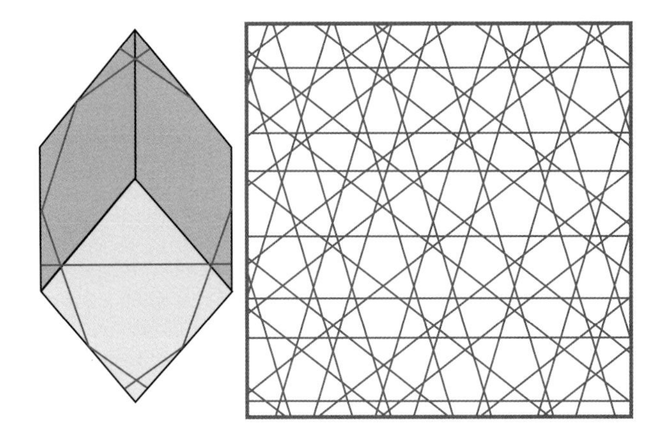

Figure 10.8
Matching Amman lines drawn across the tiles confirm that they are non-periodic, and reveal an underlying quasi-periodic pattern in 1-dimension: the Fibonacci sequence and Golden Mean.

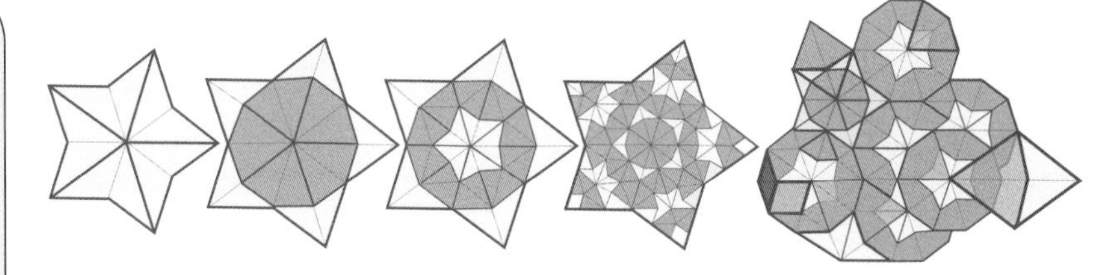

Figure 10.9
The self-similarity of Penrose tilings at different size scales showing darts and kites, and superimposed rhomboids from Figure 10.7 (re-drawn with modifications from Tovrstra, Wikipedia).

sets of parallel lines spaced in long and short intervals that relate to numbers in the Fibonacci sequence and are oriented normal to the five axes of the regular pentagon (Figure 10.5). Penrose tilings thus generate patterns that are governed by the Golden Mean and are *quasi-periodic* due to this underlying long-range order in one-dimension (1-D).

Parts of this two-dimensional Penrose tiling pattern do, however, recur at different scales so that any particular patch within the tiling is contained somewhere inside every other, and is self-similar. This hierarchical pattern continues indefinitely (through inflation and deflation symmetry) as a true fractal (Ramachandrarao et al., 2000) (Figure 10.9).

The fractal

A fractal can seem like a rather unusual concept (at first) because its dimensions are *between* 1 and 2, or *between* 2 and 3, etc. (Mandelbrot, 1983; Weibel, 2009), but leaving the mathematics aside, this self-similar principle also has a fundamental simplicity that was recognised by both Buckminster Fuller (1975, p. 266) and the artist M.C. Escher (Ernst, 1985); and describes the hierarchies of intermediate filaments within the cytoskeleton (Qin et al., 2009); spectrin (Figure 6.4, page 50); collagen (Martin et al., 2000) (Figure 6.7, page 52); tissue growth (Figure 3.9, page 30); myofascia (Turrina et al., 2013) (Figure 6.11, page 55); cranial sutures (Yu et al., 2003) (Figure 8.3a, page 71), the mammalian lung (Weibel, 2009) and tensegrity models (Figure 3.7, page 30), etc. and is then referred to as fractal-like (Jelinek et al., 2006). Thus the appearance of similar shapes at multiple size scales is obviously more than just coincidence.

The connecting links

Two further shapes now become of interest from a biological perspective because both are directly linked with the multiple symmetries of the icosahedron: the icosidodecahedron with its twenty triangles and twelve pentagonal faces, and its dual the rhombic triacontahedron with its thirty rhombic faces (Fuller, 1975, 717.01; Twarock, 2006; Edmondson, 2007, p. 58) (Figure 10.10a & b). It is further interesting that the diagonals that cross each of the faces of the rhombic triacontahedron are related to each other through the Golden Mean. It is thus the 5-fold and icosahedral symmetries of all these shapes, the Penrose tilings and quasi-equivalence principle that now enables all those previously unclassified viruses to be placed in a geometric system that includes *all* the 'spherical' viruses (Twarock, 2006) (Figure 10.10c), and it is the Golden Mean that unites them; with the capsid proteins self-assembling into this particular arrangement because it is the most energy-efficient configuration.

Spherical geometry thus shows how the 'straight' lines of a 2-dimensional surface can be curved (similar to the lines of longitude on a sphere), and that the Penrose tilings and Golden Mean are a more efficient way of dividing it up. However, it should be recognised that this is a huge simplification of a complex and developing subject. What we are ultimately trying to do is understand a bit more about the underlying geometry that underpins biological structure, and explain how biotensegrity links them together. Although icosahedra cannot close-pack to fill 3-D space (E^3) completely, the Golden Mean and space-filling rhombic dodecahedron (Figure 10.2) link it to another class of minimal-energy structures that can.

Quasicrystals

Quasicrystals (Lord, 1991) are ordered structures that continuously fill 3-D space but lack translational symmetry, like Penrose tilings, with each unit part having a different arrangement surrounding it. Quasicrystals with multiple symmetries are recognised in metallic alloys (Lord & Ranganathan, 2001b),

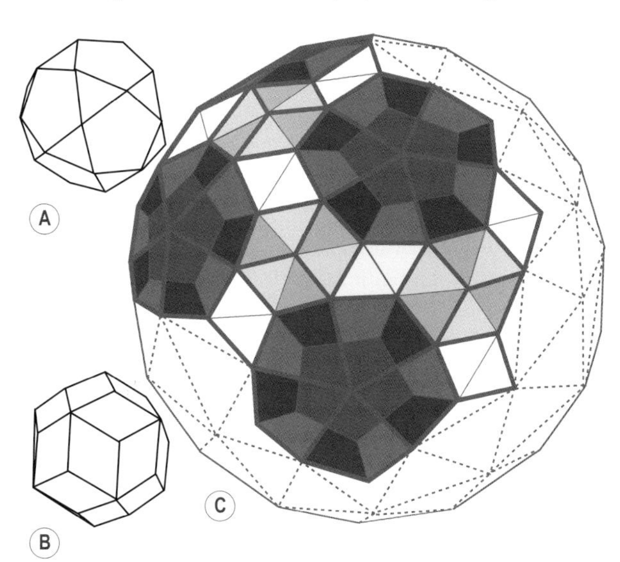

Figure 10.10
The symmetries of the A icosidodecahedron and B its dual the rhombic triacontahedron underlie C the quasi-equivalent tiling of kites and rhombs that represent the close-packing arrangement of capsid proteins overlaying the icosahedral template of Polyomavirus (re-drawn with modifications from Twarock, 2006).

Figure 10.11
A representation of tropocollagen with the three slightly contorted (quasi) tetrahelixes of procollagen (Figure 6.2b, page 50) surrounding the regular tetrahelical core (Figure 6.8, page 53).

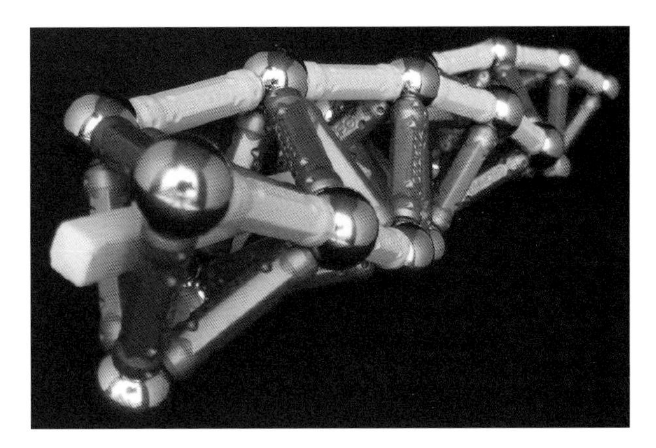

different phases of water (Johnstone et al., 2010), 'liquid' crystals (Tschierske et al., 2011), micelles (Fischer et al., 2011) and cell membranes (Palestini et al., 2011). The Golden Mean also relates them to such things as the pattern on flower heads (Figure 10.3) and tropocollagen, with its three procollagens that enclose a tetrahelical core (Sadoc & Rivier, 2000; Martin et al., 2000) (Figure 10.11).

In fact, quasi-crystals can have *any* symmetry in *any* number of dimensions, and any quasi-periodic pattern of points can be formed from a periodic pattern in some higher dimension, which suggests that all highly organised biological structures may be similar; but to better appreciate the significance of this multi-dimensional world, we must look at some examples.

Higher dimensions

We all know that the earth is round and that the spherical surface of a globe (S^2) can be projected onto a flat two-dimensional Euclidean surface (E^2), and even though the map is now distorted, the details are still recognizable. Similarly, when we look at a flat drawing in E^2, our brains make sense of the image by mentally visualising it in E^3 (3-D), because we are used to seeing things in this way, but several images are sometimes needed in order to give a good representation (Figure 10.12). In the same way, four-dimensional objects (E^4) can be projected into E^3, but this time we don't recognise them easily because this higher dimension cannot be visualised directly. The same thing happens with even higher dimensions such as 5-D and 6-D.

Penrose tilings are really two-dimensional quasicrystals obtained from a *regular* periodic lattice in five-dimensions, by projection, while those that describe the

Figure 10.12
Different views of a three-dimensional cube projected onto a two-dimensional (planar) space.

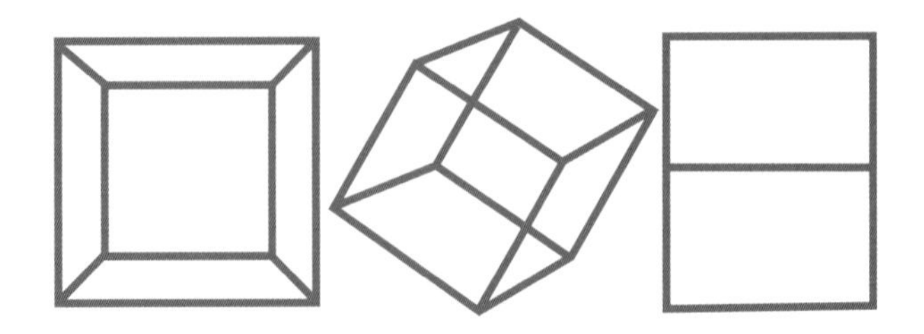

icosahedral viral capsids (Figure 10.10) are obtained from a suitable lattice in 6-D (Twarock, 2006). Within tropocollagen (Figure 10.11), the quasi-tetrahelixes of procollagen also form a contorted stack of icosahedra when viewed from our usual 3-dimensional perspective (E^3), but these become perfectly regular when projected into E^4 (Sadoc & Rivier, 2000; Lord & Ranganathan, 2001a).

Quasicrystals link simple geometry with molecules (Figure 10.11), viruses (Figure 10.10) and more complex tissues, and reveal how geometric shapes in one particular dimension can look different in another, and how structures that appear to be disorganised in our everyday world may be better understood at some higher mathematical dimension (Figure 10.13). To emphasise the point, tetrahelixes are unable to fill 3-D space (E^3) completely because gaps are left in between them, but they do close-pack in four dimensions (E^4) by curving around the surface of a hypersphere (S^3) and forming a regular icosahedron and torus (Castle et al., 2012). This takes us nicely into *hyperbolic* geometry.

Hyperbolic geometry

Hyperbolic geometry is the last of our different geometries, and generally the least appreciated, but what it has in common with the others is its two principle curvatures (Castle et al., 2012). In spherical geometry (S^2) these are both positive, and the surface bends equally on both sides; in Euclidean geometry (E^2), both of these curvatures are zero and result in a flat plane; but in hyperbolic geometry (H^2) one is positive and the other negative, which can give the surface a saddle shape such as the catenoid, etc. (Figure 10.14) (Terrones et al., 2001) but it also gives rise to other models such as the Poincaré disc and Klein bottle.

While curves with constant curvature and torsion become helixes (Chouaib et al., 2006), hyperbolic geometry relates helical tensegrities to the catenoid because it is also a minimal-energy shape (Isenberg, 1992, p.12; Motro, 2003, p. 58) (Figure 10.14). The Poincaré disc also produces remarkably organic-like patterns (Castle et al., 2012) (Figure 10.15). The Klein bottle (Figure 10.16)

Figure 10.13
The 5-fold and other symmetries of the icosahedron commonly appear in animals and plants, and whose tissues may also be related to quasicrystals.

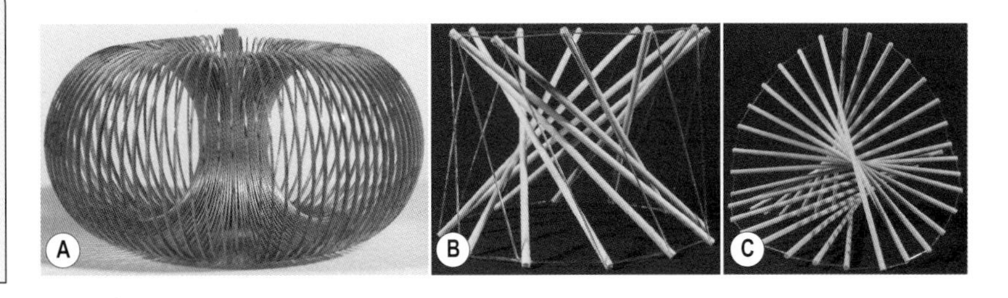

Figure 10.14
Hyperbolic geometry displayed in: A the central catenoid within a torus; B and C a T12-prism and other tensegrity configuration © Rory James.

is different altogether because it is described as a four-dimensional object that *cannot* be visualised directly, and only has one surface with no distinct boundary between inside and outside (similar to a Möbius strip); the 'bottom' of the bottle curves around and re-enters itself *without* penetrating the surface, before opening out at the 'top'.

Tensegrities can then be compared with Klein bottles because they also have no distinct boundaries between inside and out, and are able to fold and interpenetrate themselves in complex ways (Fuller, 1975, 783.00). The fundamental mathematics of both the Klein bottle and tensegrity have also been related to embryonic development and the genetic code (Rapoport, 2011). Anatomical structures commonly blend with each other so that there is no real boundary between them (Guimberteau, 2005), and the interpenetration of fascial sheets through different tissues (Figures 6.11 and 6.12, pages 55 and 56) suggests that their organization may be similar to the Möbius band and tensegrity Klein bottle (Levin, 2014) (Figure 10.17). Hyperbolic and spherical geometries thus reveal a definable order that was once unrecognised, and they blend seamlessly with the concept of biotensegrity.

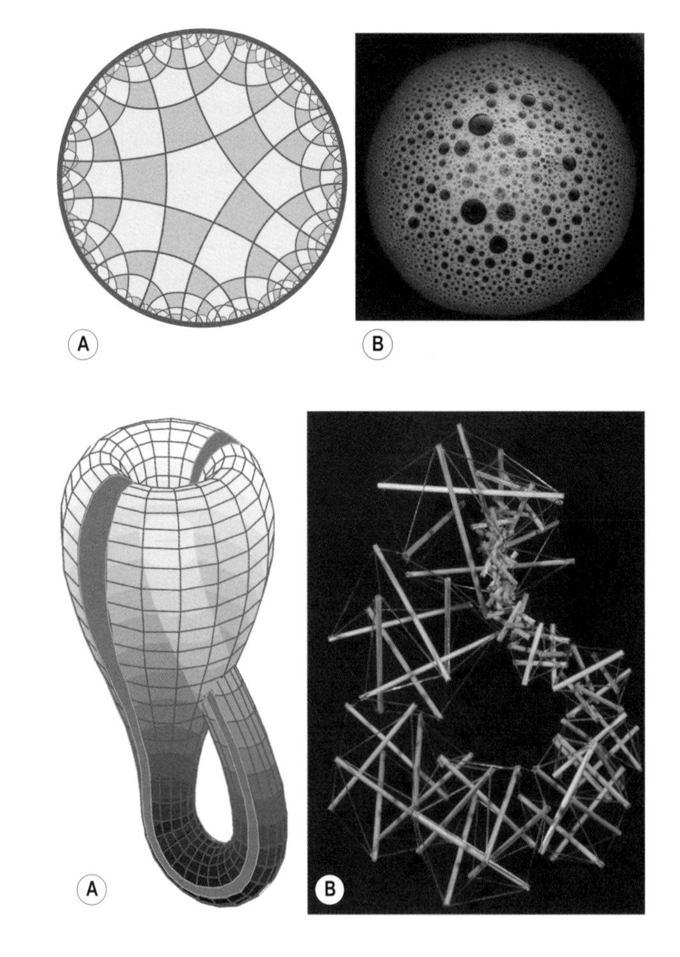

Figure 10.15
Poincaré disc based on: A squares and pentagons; and B bubbles on the surface of a stirred cup of coffee that spontaneously form an approximation of this pattern because of the interacting forces between them.

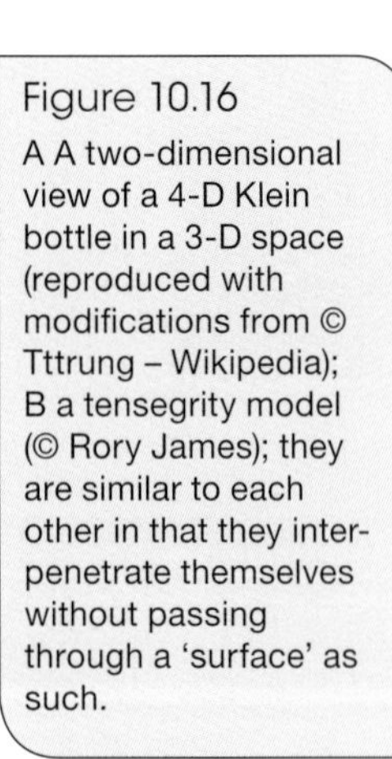

Figure 10.16
A A two-dimensional view of a 4-D Klein bottle in a 3-D space (reproduced with modifications from © Tttrung – Wikipedia); B a tensegrity model (© Rory James); they are similar to each other in that they inter-penetrate themselves without passing through a 'surface' as such.

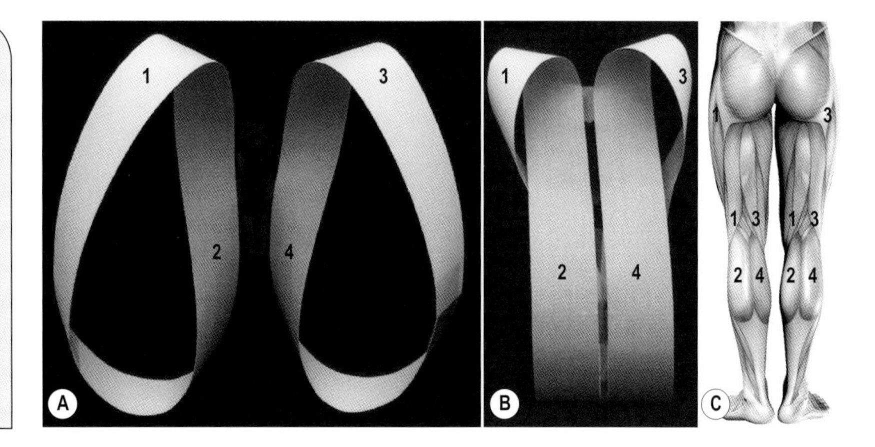

Figure 10.17
A the two Möbius strips on opposite 'sides' of a Klein bottle (the shaded longitudinal lines in Figure 10.16a) twist around so that they only have one surface; B the same arrangement showing the comparison with muscles in the lower limb C (layout courtesy of Craig Nevin).

What does this all mean?

We have looked at many geometric shapes during the course of this book, and revealed how their formation is the consequence of interactions between some basic rules of physics. Structural biology is also influenced by the same principles, which is why similar shapes and patterns frequently appear at multiple scales, but order and regularity is often hidden within them.

Until relatively recently, the world was described solely within the three-dimensions of Euclidean space and the laws of classical mechanics, and it was inevitable that this restricted view would have coloured our understanding of biology. Spherical and hyperbolic geometries, the Golden Mean, Penrose tilings, quasi-crystals and multiple dimensions, etc. are now revealing a definable order in anatomical structures that were once thought to be too complicated to understand, and biotensegrity is inextricably linked with all of this because it is based on the same principles.

The principles of biotensegrity are the same at every size scale, from the smallest molecule to the largest organism, which means that every part of a complex hierarchy can be related to the overall structure and its functions. All the examples in this book are examined from this perspective. The penultimate chapter now re-examines human anatomy from a functional perspective and reveals that there are many more orthodoxies that now need to be changed.

11

Biotensegrity: a rational approach to biomechanics

...the mere looking at externals is a matter for clowns, but the intuition of internals is a secret which belongs to physicians. Paracelsus (Waite, A.E. ed., 1894)

Biomechanics is concerned with the structural mechanics of biology and seeks to understand how different anatomical parts are able to interact in ways that cause motion. However, while our knowledge of anatomy and biomechanics has been built up over many centuries, we have already seen that levers and classical mechanics are inadequate in explaining biological movement, and that geometric patterns and shapes play a more important part in an organism's development than has been generally recognized. In a similar way, some anatomical dogmas that have been taken for granted can now appear quite different in the light of new research.

The skeleton

In the reductionist world of classical anatomy, muscles, connective tissues and bones are all described as distinct entities, in both name and function. Bones form the 'skeletal' system, muscles form the 'muscle' system and fascia is part of the 'connective tissue' system. However, during embryological development all these tissues are derived from the same mesenchymal stem cell population and differentiate within it to form a continuum that starts life as a soft-bodied skeleton.

We usually think of the term 'skeleton' as a hard bony endo- or exo-skeleton; but this is far too limiting as it actually means *"the essential framework of a structure"* (McLeod 1987, p. 935). In soft-bodied creatures and those with soft parts, such as octopus tentacles, the elephant's trunk and mammalian tongue, a helical fibrous network contributes to this framework and provides skeletal support (Kier & Smith, 1985) (Figure 6.10c, page 54). Even the vertebrate foetus must have a tensioned framework if it is to resist being crushed by the uterine wall because at this early stage it resembles the soft-bodied creatures from which the organism probably evolved (Henderson & Carter, 2002; Levin, 2006).

Bones

As the embryo grows, a soft cartilaginous skeleton develops within this fibrous network and changes its structural characteristics; and in the rudiments of long bones, growing populations of cartilage cells push on surrounding perichondrial cells and cause them to become tensioned and elongated as they form tubes surrounding the developing bone (Henderson & Carter, 2002;

Blechschmidt, 2004, p. 61). Osteoblasts then contribute to the laying down of hydroxyapatite and form the bony skeleton, with the crystals *integrated* into the fibrous matrix and forming a stiff composite material (Olszta et al., 2007). Some of the cranial vault bones form differently by condensing *inside* the fibrous membrane surrounding the brain, and separating it into an outer periosteum and inner dura mater membrane during growth (Figure 8.4, page 72) (Pritchard et al., 1956). In both cases, it is the coupling between chemical signals and mechanical loading of the fibrous matrix that induces cells to form cartilage and/or bone (Chen & Ingber, 1999; Rubin et al., 2006), and this is now described through biotensegrity (Ingber, 2008) (Figures 5.5, page 47 and 8.5, page 72).

Muscles

The muscular system also develops from the same embryonic mesenchyme, with myogenic precursors differentiating into myotubes that segregate into distinct muscles, and they have their *own* skeleton (Kardon et al., 2003). Individual muscle fibres are enclosed within a delicate network of endomysium and form part of a tensioned myofascial network that continues with fascicular perimysium and epimysium covering the entire muscle (Turrina et al., 2013) (Figure 6.11, page 55). These skeletal sheaths then form helically reinforced compartments or 'tubes within tubes' that maintain their integrity and translate the contraction of muscle fibres into bone movement entirely through tendons – or so it was thought until quite recently. Most muscles also send myofascial fibres and muscular slips to other connective tissues such as aponeuroses, ligaments, capsules and fascia, etc. (Stecco et al., 2007a), and transfer a significant proportion of their force through them.

Connective tissues

Textbooks have always attempted to fit anatomy into a neat classification system, where connective tissues are divided into 'special' types such as bone and cartilage; haemolymphoid tissues such as blood cells and lymphoid tissues; and 'general' types of connective tissues. At the same time, the 'structural' tissues are defined as those consisting predominantly of a fibrous extracellular matrix, which is itself secreted by the cells within it, and the composition of this determines another division into bones, tendons, ligaments, fascia and other connective tissues (Standring, 2005, p. 36). Bones are then considered to form the 'skeleton', tendons and ligaments have ancillary functions, and the rest are generally lumped together in order to tidy things up.

The fascia

The fascia is a fibrous connective tissue that typically refers to planar sheets that surround and interpenetrate bones, muscles, organs, nerves and blood vessels; but it is also continuous with joint capsules, muscular septa, ligaments, retinacula, aponeuroses, tendons and other collagenous tissues. Although once dismissed as a packing tissue of little consequence, it is now recognised that the fascia is a continuous uninterrupted web of tissue that links the whole body into a single tensioned network, and that extends from head to toe, from front to back and from the skin to the deep insides (Schleip et al., 2012). The fascia maintains structural integrity, provides support and protection, and acts as a shock absorber. It also plays an essential role in hemodynamic and biochemical processes; a system for intercellular communication (Langevin, 2006); and after injury, creates an environment for tissue repair. Impressive stuff!

The microvacuolar system

At an even smaller hierarchical level, collagen and elastin fibres form a tensioned network that is continuous with bones, tendons, fascia, blood vessels and nerves, etc. and links them all together as the most basic level of tissue organization (Guimberteau, 2006; Kapandji, 2012) (Figure 11.1). These fibres enclose polyhedral microvacuoles, or 'bubbles' filled with a slightly viscous 'jelly' containing water, fat and hydrophilic proteoglycans that are constantly changing shape as the fibres adjust their position in response to the movement of surrounding tissues. This fractal-like system then allows adjacent structures to slide in relation to each other during movement, dampens the transfer of excessive forces and protects the connecting nerves and blood vessels embedded within it.

It is thus apparent that all these different skeletal tissues blend into each other and that there is no clear distinction between them at the micro level. As they all have a common origin in cells of the embryonic mesenchyme, they should no longer be considered as separate systems but mutually dependent specializations of the same mesodermal tissues.

Figure 11.1
The microvacuolar system consists of polyhedral 'bubbles' made from collagen and elastin fibrils surrounded by a proteoglycan rich gel, with the fibre 'walls' connecting, separating and sliding in relation to each other during movement. (re-drawn from Guimberteau, 2005).

A new reality

Mesokinetics

Unfortunately, and for perfectly understandable reasons, textbooks have reduced human anatomy down into a collection of distinct 'parts' and these are then re-grouped together into functional units. Bones and 'skeletal' muscle, for example, are considered together as the 'musculo-skeletal' system, and they do occupy a substantial proportion of body weight (15% and 40% respectively) (Tanaka & Kawamura, date unknown) but the imagery associated with this term has relegated the other connective tissues to a mere supportive role that is no longer justified.

A new name, the *mesokinetic* system, is a term that is much more inclusive and also describes their origins and functions (Levin & Scarr, 2012); the *meso* as in mesoderm is the embryonic tissue layer from which the connective tissues are all derived, and *kinetics* is the branch of mechanics concerned with the study of bodies in motion (McLeod, 1987, p. 554).

The introduction of this new name changes the outdated and misleading perception of a muscle/bone duality that has dominated for too long, and includes anatomy at every scale from the cytoskeleton upwards. If the whole of the mesokinetic system is united in the common purpose of structural integrity and movement, this term must also conform to the definition of an organ: *"a part of the body... adapted and specialized for the performance of a particular function"* (Walker, 1991, p. 632). The mesokinetic system as the 'organ of movement' then becomes the inclusive structural and functional entity upon which the biomechanics of tensegrity can now proceed; but first we must unpick some of the other assumed conventions.

Unravelling the old ideas

It has always been assumed that tendons and ligaments transfer tensional forces across synovial joints in a parallel system, i.e. side by side, with tendons transferring the force of muscle contraction to bones, and ligaments acting as mere guide wires and restrainers of excess joint motion; but new research suggests otherwise. For example, most muscles in the lateral elbow region transmit their force through strong intermuscular connective tissue layers, and these converge on the humerus as the common extensor 'tendon' (Figure 7.10, page 66), but the myofascia that permeates and surrounds them is also connected to the humerus, ulna and radius through ligaments, and is continuous with the deep fascia that links multiple muscles together.

A misplaced wisdom

Traditionally, the fascial tissues surrounding muscle are removed during dissection, on the implicit assumption that their influence is negligible, but a procedure that removed muscle fibres rather than connective tissues found that the ligaments in the elbow could not be distinguished as distinct entities and were actually artifacts of the dissection process. The orientation of most collagenous fibres was also found to occur between bone tissue and muscle fascicles, rather than bone to bone as would be expected with a passive restraining ligament. The position of proprioceptive sensors was similarly organised according to the functional architecture of muscles arranged in *series* (end to end) with dense connective tissue layers such as ligaments and aponeuroses, challenging the usual distinction between muscular and articular mechanoreceptors (Van der Wal, 2009).

In a similar way, experiments (in rats) that used muscles with associated fascia intact showed that the forces appearing at both ends of a stimulated muscle differed from each other and varied according to the relative lengths of *adjacent* muscles; and that instead of the contractile force being transferred entirely through tendons, it was also being distributed to surrounding muscles and connective tissues through associated myofascial sheaths (Huijing & Baan, 2003). These anatomical continuities between different muscles then enable reciprocal feedback to occur over multiple pathways, both mechanical and neural, with a variety of proprioceptors in the connective tissues monitoring the precise timing, intensity, duration and release of tissue deformation (Stecco et al., 2007b). As length changes of poly-articular muscles are determined across two or more joints, they are more likely to change their relative positions with respect to mono-articular muscles, and alter the balance of tension in associated fascia as the joint angle changes.

A global synergy

It is now quite apparent that multiple connective tissues are involved in the transmission and distribution of tension, and that the functional distinction between tendons, ligaments, aponeuroses and fascia, etc. is no longer a clear-cut relationship. Even the attachment of muscle spindles to the perimysium suggests that they are monitoring connective tissue tension rather than muscle activity *per se* (Stecco, 2004).

The fascia profunda, or 'deep' fascia, is a particular tissue within the limbs that is intimately connected to most muscles in the region and receives a significant number of fibrous and muscular expansions from them (Stecco et al., 2007a). As a consequence, these fascial continuities influence the amount of tension produced by both synergist (Huijing & Baan, 2003) and antagonist muscles (Huijing & Baan, 2008) and, if so-called 'antagonists' can work as a co-operative synergy in controlling joint movement, such as brachioradialis and anconeus in the elbow (Scarr, 2012) (Figure 7.10, page 66), the whole concept of flexor/extensor and agonist/antagonist then becomes highly questionable (Levin, 2002; Turvey, 2007). Any part of a synergistic relationship can also have multiple roles in relation to other synergies, such as brachioradialis with biceps in the distal radio-ulna joint (Garcia-Elias et al., 2008) which means that their ability to function within each relationship must depend on a huge number of mechanical and neural interactions.

The simple complexity of motion

Biological structures are dynamic, with a constant turnover of cells that are growing, differentiating, reproducing and dying, and changing the mechanical behaviour of every tissue over time (Henderson & Carter, 2002; Blechschmidt,

2004). In addition, the cytoskeleton (Ingber, 2003a), myofibroblasts (Schleip et al., 2006) and muscle cells (Masi & Hannon, 2009) are continually generating tension that must be counterbalanced by other tissues, if stability is to be preserved, which is why the stress/strain relationship in *living* tissues never reaches zero (Levin, 2009) (Figure 4.1b, page 34).

Traditionally, muscle activity has been reduced to a spurious state of 'on' or 'off' because electromyography is unable to measure this basal tone. Ultimately, it is this state of intrinsic tension (pre-stress) that removes any slack and primes the system for action, as well as enabling connective tissues to sense and respond to changes in mechanical stresses from virtually anywhere in the system (Levin, 2006; Turvey, 2007). As a result, changes in tension can cause a cascade of neural signals to be transmitted directly to the muscles and initiate a change in their activity, or combine with those from the central nervous system and cause a more directed response from higher centres.

Muscles should thus no longer be considered as the 'motors of movement' but dynamic regulators of connective tissue tension (Levin, 1982; Van der Wal, 2009), with reciprocal interactions between all these different mesokinetic tissues within a tensegrity configuration providing the most energy efficient platform for a joint to move (Turvey, 2007).

The dynamics of movement

Movement is a characteristic feature of animals and the means that they use to express life; whether walking, talking, eating, playing, laughing or crying, the whole of an organism's being is impelled towards it – even the cytoskeleton and visceral 'powerhouses' depend on movement if they are to function properly. Although biomechanics has traditionally described this through a system of isolated levers (Figure 4.3, page 37), we have already seen that this is a poor representation of movement and that biotensegrity overcomes many of its inconsistencies. Synovial joints are really specialised regions within a global tension system that enables the structure itself to guide movement, and muscles to provide the power and moment-by-moment adjustments at a higher level of control.

The control of motion

Movement is the change in position from one place to another, and must be carried out in a controlled way if it is to be purposeful. Closed-chain kinematics describes the geometry of how biological structures can do this. It couples multiple 'parts' into a continuous mechanical loop, with the motion of each 'part' causing controlled changes in position, velocity and kinetic energy in all the others and, as tensegrity configurations, allows them to fill space in every direction and create an energy-efficient control strategy that is built into the very structure itself (Levin, 2014) (Figure 4.9, page 41).

The simple tensegrity-icosahedron demonstrates this with its 'jitterbug' contraction and expansion (Figure 11.2a), and a similar model with two struts

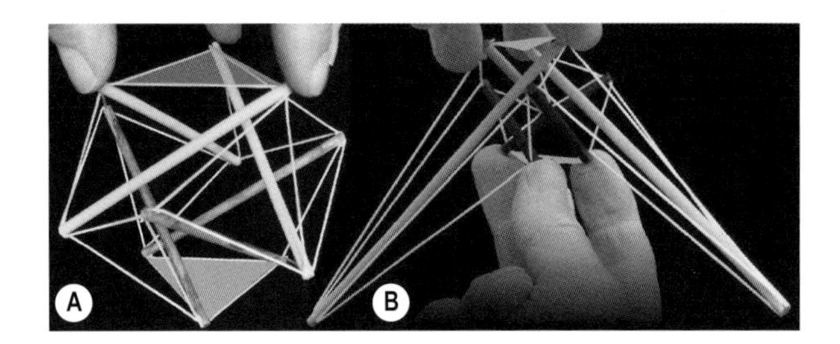

Figure 11.2 A
A tensegrity-icosahedron showing two chiral groups of struts, and that pushing the 'tension-triangles' together causes the equatorial nodes to twist and the entire structure to contract; B a similarly configured model with two struts elongated to represent the bones in a 'joint' shows that twisting the 'tension-triangles' simulates a limited 'flexion' and 'extension'.

elongated to represent bones on either side of a 'joint' (Figure 11.2b) shows how the ends of these struts move towards each other, and back again, when the 'tension-triangles' are twisted around a common axis. Whilst we might compare this simple motion with 'flexion and extension', it is really a *realignment* of the 'bones' that are responding to the structural constraints within the system (Levin & Martin, 2012). As muscles and connective tissues provide tensional continuity across multiple joints, the notion of a synovial joint as a distinct entity then becomes rather quaint.

Functional kinematics

The functional value of multi-bar kinematic chains has now been described in the jaws of fish (Roos et al., 2009), reptiles (Kardong, 2003), birds (Hedrick et al., 2011), and the limbs of mammals (Usherwood et al., 2007). In humans, the bones, muscles and connective tissues of the shoulder, elbow (Figure 7.9, page 65), knee (Figure 7.7, page 63) and pelvis/spine/hip complex, for example, all become tensioned bars that guide motion (Levin, 2006), and muscle bars that span multiple joints result in energy-saving power transfers (Biewener, 1998). Even joints with no muscles crossing them such as in the limbs of large terrestrial animals (Van den Bogert, 2003) (Figure 11.3b), or where there is an insufficient number of muscles to account for the complexity of movement (as in human fingers) (Figure 11.3a) (Valero-Cuevas et al., 2007), stability and control must inevitably depend on the geometry of the connective tissues.

This outsourcing of complex control tasks directly into the structure efficiently reduces the amount of effort needed to coordinate changes during movement (also a characteristic of tensegrity) and allows the motion of 'joints' to be guided by the particular configuration of soft tissues surrounding them. For example, it has been suggested that the particular tendon configuration in the fingers acts like a logic gate that switches joint position (Valero-Cuevas et al., 2007) (Figure 11.3a), with the anatomy functioning as a 'dynamic information processing network' (Ingber, 2003b) that is able to perform computations normally attributed to the nervous system. The intrinsic tissue tension and coupling of muscle cells to the nervous system then means they can respond quickly to proprioceptive signals (via spinal reflexes), and make any further adjustments necessary to maintain stability, with the central nervous system exerting control at an even higher level.

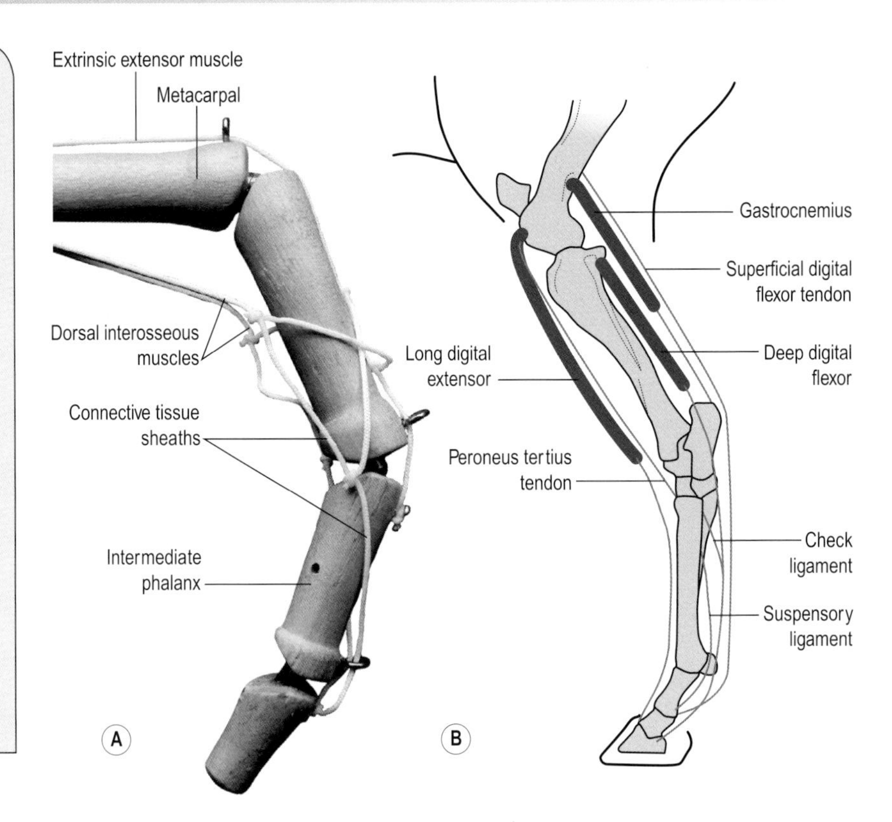

Figure 11.3
Closed-chain kinematics: A a model of the metacarpal and phalanges of the middle finger showing the simple geometric configuration of muscles and connective tissue sheets that has been proposed to regulate the complex combinations of joint movements (Valero-Cuevas et al, 2007); B joint movement in the distal hind leg of a horse is controlled entirely by the closed-chain kinematic geometry of polyarticular muscles and connective tissues (re-drawn from van den Bogert, 2003).

A shift in balance

Whilst the function of each part relies to some extent on all the others, any 'aberrant' mechanical signal has the potential to compromise it, and a genetic mutation that leads to a change in the shape of a particular molecule could alter its structural relationship to others and lead to pathology (Ingber, 2003c; DuFort et al., 2011). Developmental abnormalities, postural misuse and injury to tissues in one region will also cause changes in the tensional balance of others some distance away, and might jeopardise their functionality as they adjust to a different structural configuration – as was described with the cranial vault, etc. The effect of an acute muscle contracture on surrounding tissues is often clear to see as the body contorts itself into a position that minimises the local stress but compromises the bigger picture, at least temporarily.

Therapeutics

The ability of the human body to function in the way that it does and repair itself when things go awry has been a source of wonder for millennia, and an incalculable number of therapeutic interventions have been devised to assist these processes. However, while an appreciation of the fascial network in relation to health and disease has been an integral part of manual therapy for more than a century (Still, 1899), the concept of biotensegrity has been around for a much shorter time (Robbie, 1977; Levin, 1981; Ingber, 1981), and orthodox views of anatomy and biomechanics have continued to dominate. Although (bio)tensegrity is frequently mentioned in the literature, its precise value to clinical practice is often unclear. Is it simply a footnote to technique or does its influence go beyond theory; and is it possible to see or feel the effects of tensegrity in the body?

A change in perception

Models have always played an important part in clinical practice and provide the practitioner with a conceptual tool that can be used in both diagnostic and therapeutic reasoning (Hohenschurtz-Schmidt, 2013). Levin (1981) noticed that normal bones do not touch each other but appear to float within the soft tissues and that tensegrity provided a plausible explanation for these findings; and every hands-on practitioner recognises the balance of tension and compression in the body. The mesokinetic system as a tensegrity configuration then becomes the unifying structure that contains bones and muscles, etc. *within it*, rather than as separate entities, with each performing its own localised functions in relation to whole-body physiology. Joints are then just particular anatomical regions that provide the flexibility necessary for movement and are inextricably linked with each other (Stecco, 2004; Myers, 2014), and the entire kinematic system working as a whole allows them all to function the same in virtually any position (Levin, 2002, 2014) (Figure 11.4).

Figure 11.4
Cuban acrobatic artists Leosvel and Diosmani demonstrate that the human body can function much the same in any position. Reproduced from © Ludovic Péron, Wikipedia.

The resolution of a 'local' condition can then require a 'whole-body' approach to treating it (or vice versa), particularly if tissues some distance away have become adapted to changes in the overall structural balance; and an understanding of biotensegrity provides the rationale for this (Kassolik et al., 2013; Schleip & Müller, 2013; Martin, 2014). It recognizes that removing the restrictions to normal function, and changing the geometry, is part of what is necessary to allow the body's inherent self-balancing mechanisms to restore homeostasis.

The biotensegrity model

Time spent with the 'stick and string' models shows how every part of a structure can be integrated into a functioning mechanical unit through reciprocal tension, and that their fluid-like dynamics are quite similar to those observed in living tissues, with the particular quality of motion sometimes described as 'tensegral'. They demonstrate the point of 'balanced tissue tension' that is observable in practice; and how 'disturbances' due to posture, trauma and pathology might alter this. The harmonic coupling between different parts enables them to communicate mechanically and particular movement patterns to be 'memorised', and may even explain the palpable findings that have been interpreted as 'tissue memory' (Upledger & Vredevoogd, 1983; Oschman, 2012). Such models reveal the forces that are active within them, the basic geometry that underlies all natural structures and the functional mechanics that develop from it. Treatment is then more than just changing tissue tensions, repairing the structure or improving mobility, but a process that alters the tissue geometry (Levin, 2014); and this then allows the body's homeostatic mechanisms (functional biotensegrity) to re-establish a new position of dynamic equilibrium and improve function.

A cautionary note

It is important to note, however, that tensegrity is *not* a particular type of treatment (although sometimes described as such), and there are no 'tensegrity techniques' (Scarr, 2014). Manual therapists have long considered the body to be a functionally integrated unit (Still, 1899; Findley & Shalwala, 2013), even before tensegrity was discovered, and an innumerable number of successful treatments based on the old lever system have been performed. The value of the tensegrity model is *not* that it necessarily changes a particular approach to treatment but that it provides *a better means to visualise the mechanics of the body* in the light of new understandings about functional anatomy.

Basic science

All models are representations of a more complex reality, and their acceptance is not dependent on the piecemeal accumulation of scientific 'data' but because they describe the dynamic behaviour of biology better than any other. The biotensegrity concept reduces structure to its simplest form and is based on fundamental physical principles rather than artificially contrived ones. This is what makes tensegrity a part of basic science, and an understanding of this should underpin what every clinician does!

12
Biotensegrity: the structural basis of life

Biotensegrity is a structural design principle in biology that describes a relationship between every part of an organism and the mechanical system that integrates them into a complete functional unit. It is a simple re-evaluation of anatomy as a network of structures under tension and others that are compressed; parts that pull things together and others that keep them apart – basic physics!
Graham Scarr

A biotensegrity configuration is also an information processing network (Ingber, 2003b) that combines complex anatomy and biomechanics as manifestations of a dynamic energy system, and potentially transfers harmonic oscillations from the cell nucleus to the entire mesokinetic system (Pienta & Coffey, 1991; Alippi et al., 2010). While mathematical models confirm its significance to the cytoskeleton (Baudriller et al., 2006; Chen et al., 2010), multi-cellular systems (Jamali et al, 2010) and larger structures (Lobo & Vico, 2010), it is also attracting the interest of engineers who devise novel ways to construct buildings (Jauregui, 2010), bridges (Rhode-Barbarigos et al, 2010) and robots (Rieffel et al., 2010; Moored et al., 2011; Orki et al., 2012), and with ideas that feed back to further biological research. The brief foray into different types of geometry and multiple dimensions (Chapter 10) suggests even more lines of enquiry.

In the simple model, every component contributes to stability of the whole structure and has a mechanical influence on every other. The same principle applies to living systems, where the vast number of connections and variations in the mechanical properties of different tissues can seem impossibly complicated, and are difficult to analyse (using conventional methods) (Lin et al., 2006). Biotensegrity naturally simplifies things by having just two components, tensioned cables and compressed struts, with the stick-and-string models demonstrating the mechanics of living structure better than any other. Simple geometric shapes and the vector systems used to model the forces within them then help us to visualise biological structures as balanced energy networks and displays of the invisible forces that hold them together.

Most experimental work on biotensegrity has been carried out on cells, which can be examined individually, collectively or in parts. Single cells are easily manipulated, abundant in variation and supply, and there are few issues that might restrict their use; but multi-celled organisms such as humans are quite different. They consist of multiple hierarchical levels, with all the increases in complexity that might be expected, and examination of any part in isolation can be misleading (Figure 9.1, page 77). Cadaver experiments have their limitations because they don't give results that precisely match living structures; and *in vivo* experimentation is restricted due to practical and ethical issues. Generalisations of biotensegrity from a *macro* perspective have thus been reasoned from first principles or inferred from models and observation (Levin, 2006; Ingber, 2008).

First principles

All natural structures are the result of interactions between atomic forces and the balanced configurations that they settle into, and these forces always act in straight lines and connect over the shortest path (Figure 2.1, page 13). This means that they are geodesics, and geodesic geometry is fundamental to the formation of natural patterns and shapes because of its close-packing and minimal-energy efficiency (Figure 2.4, page 14). The attraction and repulsion between atoms causes them to form crystals and organic molecules, which then become the *physical representations of the invisible forces within them.*

The Platonic shapes and where they lead to

The ancient Greeks considered that everything in the universe could be expressed through just five Platonic shapes, a view that persisted well into the nineteenth century, but this idea was overthrown when Charles Darwin published his groundbreaking book *On the Origin of Species* (Darwin, 1859). Despite a brilliant effort by D'Arcy Thompson (1917) in relating biology to physics and geometry (Thompson, 1961), the driving force in developmental evolution came to be considered as little more than the random selection of clever artifact-like contrivances, and these just so happened to meet the 'requirements' for improved biological function (Denton et al., 2003). The genetic code and Mendelian rules of heredity then seemed to seal the fate of platonic geometry, but ironically, the discoverers of the DNA double-helix had also re-introduced it back into biology. In 1956, Francis Crick and James Watson suggested that the arrangement of proteins in the outer coating of the 'spherical' viruses was likely to have cubic symmetry (Crick & Watson, 1956), just like the Platonic shapes, and this feature was subsequently confirmed in other molecular structures (Pauling, 1964).

All spherical viruses are now known to conform to the icosahedron (Figures 10.4, page 90 and 10.10, page 93), the cube and dodecahedron appear in some enzyme complexes (Izard et al., 1999) and photonic crystals within butterfly wing scales conform to a cubic-cell Gyroid (Saranathan et al., 2010). The helix, with its geometric origins derived from the tetrahedron is a common motif in protein molecules (Figures 6.3; 6.4, page 50 and 6.7, page 52) and cell walls (Emons & Mulder, 1998). Even the functions of DNA is based on its quasi-crystalline structure and quantum-level properties (Rapoport, 2011; McFadden, 2013). Basic geometry as a major determinant of biological complexity in the sub-cellular realms has thus been reinstated (Denton et al., 2003)!

Of course, the formation of more complicated structural systems is still going to be influenced by the same laws (Fuller, 1975, 220.00), and the whole of biology is an open display of these (Levin, 2006), but using conventional methods to follow the interplay between all the forces within them is near impossible. This is why the simple stick-and-string models are so important: they demonstrate that the principles that apply at the smallest scale can also apply at the largest scale (and everything in between), *no matter how complicated they become* (Figure 12.1).

Figure 12.1
Tensegrity model showing a relationship between the icosahedron and higher-level (curved-strut) tensegrity.

The helix

We have seen how the basic helix originates with the close-packing of tetrahedra to form a tetrahelix (Figure 6.1, page 49), which then influences the dynamics of more complex structures as crossed-helical alignments of collagen in the heart, blood vessels, myofascia, intervertebral disc and entire body (Figure 6.12 and 6.13, page 56). Biological helixes are easily modelled through tensegrity (Figure 6.1c, page 49), with multiple distinct components that are tensioned or compressed within self-similar hierarchies, and that maintain fluid flow and changes in shape without buckling or collapsing (Figure 6.9, page 53), and two helical precursors (T3-prisms) of opposite chirality then combined to form another model based on the sphere and icosahedron (Figure 3.3, page 27).

The icosahedron

While the tetrahedron contained the *minimum* volume within the *maximum* surface area (Figure 2.7, page 16) so the icosahedron contains the *maximum* volume within the *minimum* surface area (Figure 2.15, page 20) and approximates a sphere better than any other shape. The flexible molecular surfaces of bubbles (Figure 2.3b, page 14), spherical viruses (Figures 10.4, page 90 and 10.10, page 93) and the cortical cytoskeleton (Figure 6.5, page 51) are all natural 'spheres' based on high-frequency icosahedra (Figure 9.8, page 85).

The tensegrity-icosahedron also serves to model the cell's internal cytoskeleton (Figure 5.2, page 44), the elbow (Figure 7.9, page 65) and cranial vault (Figure 8.2, page 70), etc. and, as a modular chain, demonstrates how the weight of a long-necked animal could be supported so far from the body without overstressing its dorsal tissues (Figure 7.4, page 59), as well as how the spinal column can function the same in both vertical and horizontal positions

(Figure 11.4, page 107). The icosahedron even links the tetrahelix with molecular quasi-crystals of tropocollagen in different geometries (Figure 10.11, page 94) and higher dimensions. It is likely to reproduce itself at multiple size scales during development and is probably ubiquitous in biological structure, which is essentially about the science of bubbles, foams and 'domes' (Fuller, 1975, 1025.10; Levin, 2006; Kapandji, 2012). The efficient sphere-like form and volume/surface area ratio of this shape give it a minimal-energy configuration that compliant organic shapes will automatically try to assume (the 'strange attractor' of natural systems) (Ingber, 2003b; Levin, 2006), even though they may be prevented from fully achieving this because of other physical constraints.

Developmental evolution

Whenever nature uses the same principle in a variety of different situations there is probably an underlying energetic advantage to its appearance, and embryonic development and evolution will automatically favour those configurations that are the most efficient in terms of stability, materials and mass. Although the genetic code plays an important part in determining an organism's particular characteristics, it can only do so if the basic elements of its structure are already in place and, it is the spontaneous attraction and repulsion between atoms that ultimately determines their development.

A living organism must remain stable at each instant of its existence, from the first embryonic cell to the full-grown adult, and its survival depends on the coherency of its construction principles (Levin, 2006). The basic rules of assembly then enable simple molecules to form polymers and more complicated patterns (Denton et al., 2002, Edwards et al., 2010) – including DNA (Liedl et al., 2010) – and become involved in the formation of larger and more complex hierarchies that are now visible (organelles, cells, tissues and organs), with each one having a direct influence on the development of all the others; and all this happens at the same time.

Within the cell, physical and chemical signals from the extra-cellular matrix and surrounding cells alter the molecular balance of the cytoskeleton, through mechanotransduction, and cause changes in cell function such as growth and differentiation, etc (Figure 5.5, page 47). Groups of cells then create tensional and compressional forces that are transmitted to other parts of the mesokinetic system and alter the larger environment (Henderson & Carter, 2002; Blechschmidt, 2004). As the cells within these tissues respond and feedback in a similar way, the resulting balance causes distinct anatomical patterns and shapes to emerge as the most energetically stable configurations (Thompson, 1961; Nelson et al., 2009).

The emergence of structure

Living structures are dynamic energy systems, and the direction and intensity of the forces within them are constantly changing, but they always follow the path of least resistance and move towards a state of equilibrium. The constant assembly and dis-assembly of cytoskeletal components (Figure 5.3, page 45), in response to these forces, then creates an interlinked and dynamic network that ultimately regulates and switches cell function. The formation of particular tissue patterns that result from this follows the same basic principles, with the self-similar fractal-like branching of bronchial/alveolar and arterial 'trees' (Figure 3.9, page 30), etc. being easily recognised examples; with the inflation of Penrose tilings (Figure 10.9, page 92), and the Golden Mean, revealing some fundamental geometric principles (Figure 10.3, page 89). The 'lost' strut model then shows that a complex system can be a much more efficient arrangement than a simple one (Figure 9.3, page 81), and is a biological imperative that drives development. It also shows how the principles of biotensegrity could enable an organism to maintain stability throughout the changes necessary to achieve this.

The survival of the fittest

This linking of every component as an inevitable part of normal development also means that stresses imposed at one level can have a direct influence on structures elsewhere in the hierarchy (because they follow the same interconnected pathways) (Figure 9.1, page 77), and can cause changes at the molecular level that stimulate remodelling of the mesokinetic tissues, according to Wolff's law, etc. Whilst these stresses create transient imbalances that are constantly being resolved in the attraction and repulsion between atoms (as part of normal function) (Gao et al., 2003; Gupta et al., 2006), their *persistence* would have disastrous consequences. The generation of unstable shear stresses (as potentially found in lever systems) would be likely to cause these structures to collapse through material fatigue at an early stage in development and that really would be the end of them (Levin, 1995). Although 'shear stresses' are commonly mentioned in the biological literature, they are mostly simplifications that have disregarded the hierarchical details of the associated anatomy.

Natural selection is an evolutionary device that enables structures that are stable to survive and unstable ones to disappear, and simply could not tolerate a precarious system of mechanics. Development is a moment-by-moment journey from simple to complex, and the survival of each part ultimately depends on the balance of physical forces within itself and in relation to its surroundings. In the natural world the most energy-efficient system *always* wins – first principles!

The biotensegrity model

From the outset, we have used tensioned 'cables' and compressed 'struts' to represent the basic elements of biological structure, and seen that this simplicity can be organised into incredibly complex hierarchies that would otherwise seem impossible to understand and, while the number of possible combinations is probably infinite, various tensegrity classifications have been introduced (Pugh, 1976; Connelly & Back, 1998; Motro, 2003; Skelton & de Oliveira, 2008).

Whilst these have not all been assessed from a biological perspective (and some are, in any case, not relevant), the simple '3-strut' remains the most basic of models (Figure 3.1, page 25). We have seen how a simple increase in the number of struts gives it the versatility to form the *cylindrical* T-helix, with its modular chain of T-prisms (Figure 3.2, page 26), and combine with its opposite chirality to form the *spherical* tensegrity-icosahedron, or T6-sphere (Figure 3.3, page 27);

and these two shapes are then used to explore even more complex structures (Figure 12.1) that can fill space completely (Fuller, 1975, 784.00).

The wheel

The bicycle wheel is one of Buckminster Fuller's simplest examples of balanced tensioned and compression but it is actually far more complex (Figure 1.9, page 8), because even though the outer rim is curved, from a biological perspective every part of the wheel is itself a hierarchical tensegrity made from smaller components (Figure 9.6, page 84). Even molecular helixes such as DNA can be compared to a wheel drawn out along the rotational axis, with the deoxyribose chain acting like the outer rim balanced by the inner nucleotide hubs.

Levin used the tensegrity-wheel to describe bones as essentially sesamoids tied in with strong muscular and fascial sheets, and that are caused to move as the tension within these tissues changes, and the strong but delicate tissues in the lungs of mammals and birds (Figure 8.8, page 75) also fit this model. Even the weight of the horse is transferred entirely through soft-tissue laminae that effectively suspend the distal phalanx (hub) from the inside of the hoof (rim) (Hickman & Humphrey, 1988, p. 53). The spider web, however, is one such network that includes the ground, tree and garden gate, etc. in its outer 'rim' and perhaps stretches the tensegrity wheel model to the limit (Connelly & Back, 1998), but at the molecular scale, the regular spacing and orientation of helical fibres and crystallite units suggest that the silk fibre really *is* a tensegrity configuration (Knight & Vollrath, 2002; Du et al., 2006) (Figure 12.2).

Multiple geometries

Structural hierarchies then reveal that curvature can be another way of minimising energy (Figure 9.6, page 84), and curves commonly occur in biology even though the forces within them always follow straight lines. The spherical geometry of the tensegrity-icosahedron demonstrates the jitterbug as an energy system that expands and contracts omnidirectionally around a central point (Figure 11.2a, page 105), and provides a direct clue to understanding the infectiveness of viruses (Figure 10.10, page 93), contraction of the heart, mechanics of the cranial vault (Figure 8.2, page 70) and non-linear dynamics of joint motion (Figure 7.9, page 65), etc, etc.

The Fibonacci sequence and Golden Mean appear in the formation of microtubules within the cytoskeleton (Figure 6.3, page 50), cells in the tips of

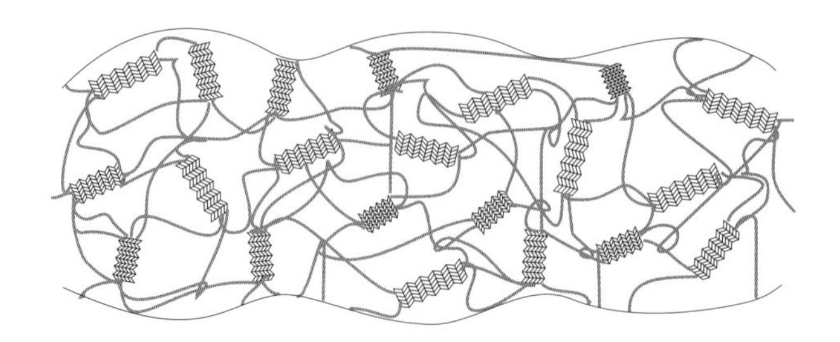

Figure 12.2
A fibril of spider silk consists of beta sheets of poly-alanine suspended in a network of helical oligopeptides (re-drawn from Termonia, 1994).

growing plants (Figure 10.3, page 89); and the close-packing of tropocollagen 'quasi-crystals' within connective tissues that pervade the body (Figure 10.11, page 94), and (maybe) collagen alignments in the outer wall and myofascia of all mammals (Figure 6.12 and 6.13, page 56). The burgeoning interest in these fundamental geometries and shapes suggests that there is much to be explored (Figure 12.3).

The hyperbolic geometry of the catenoid is an intrinsic part of even the simplest tensegrity model (Figure 10.14, page 95), and enables more complex patterns to develop as a result (Castle et al., 2012), but its potential has only just been touched on here. The tensegrity Klein bottle is an even more bizarre shape (Figure 10.16, page 96) that potentially links the interweaving of fascial sheets throughout the body with higher dimensions; and the Möbius strips demonstrate the remarkable similarities to the interweaving pattern of muscles in the leg, etc. (Levin, 2014; Nevin, 2014) (Figure 10.17, page 97).

However, in spite of all this apparent order, *"biological structures are chaotic, nonlinear, complex and unpredictable by their very nature..."* (Levin, 2006, p. 79) and a great deal of concerted effort across multiple disciplines is needed to understand them more fully. As Fuller stated *"the physical Universe is a self-regenerative process... governed by a complex code of weightless, generalised principles"* (Fuller, 1975, 220.05) – and it is worth taking notice of these principles!

The use of models to analyse the function of a particular anatomical region has been the mainstay of biomechanics for centuries (Figure 7.1, page 57), but until recently, there has been no overarching principle that could unite them all. The lever model survived as a collection of isolated parts because there was simply nothing else to replace it, but it could never have been considered complete unless all those 'parts' were functionally coupled together. The concept of biotensegrity with its closed-chain kinematics (Figure 4.9, page 41) is now such a model that intrinsically integrates *every* 'part' into a coherent whole (and consequently makes the idea of

Figure 12.3
A particular variety of cauliflower displaying the hierarchical self-similar fractal-like pattern related to the Golden Mean (Figure 10.3, page 89). Reproduced from © AVM, Wikipedia.

levers irrelevant). Although there will always be some flights of fancy (and rightly so), the potential of biotensegrity in exploring *human* biology seems to be limitless.

The 'complex' model and beyond

Tom Flemons' models of the spine and leg (Figure 3.10, page 31) provide examples of multiple joints coupled into kinematic units that can move in remarkably life-like ways (Flemons, 2012); and Gerald de Jong's dynamic computer models demonstrate the organic-like behaviour of a high-frequency tensegrity-icosahedron when dropped from a height (de Jong, 2010) (Figure 12.4).

A great deal of interest is also being shown by engineers who have taken the biotensegrity model and adapted it for use in robots that walk (Rieffel et al., 2010; de Jong, 2011; Orki et al., 2012) and swim (Moored et al., 2011), and in deployable structures in space where the combination of strength, lightness and flexibility are of paramount importance and may directly contribute to the kinematic analysis of human motion. The 'tetraspine' as one of these robots couples multiple tetrahedra into a structure that can surmount obstacles in its path, and as with the real-life version, controls movement by using multiple cables that regulate the distance between each vertex (Tietz et al., 2013) (Figure 12.5).

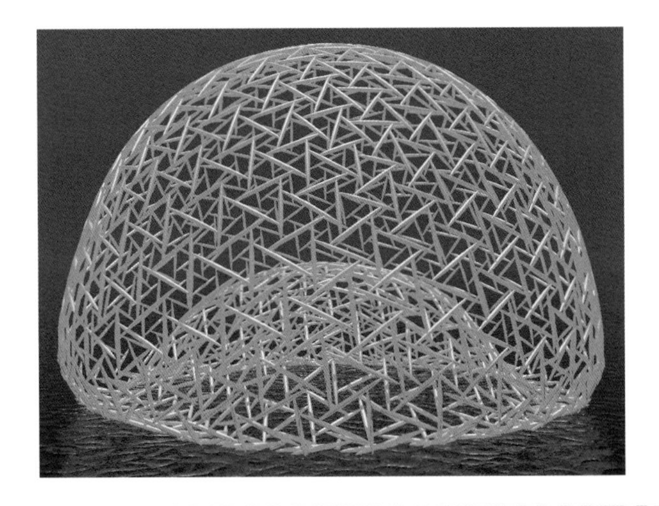

Figure 12.4
A computer model displays the organic-like deformation that occurs when a tensegrity 'sphere' (6F T-icosahedron) is dropped onto a surface. Reproduced from part of a video sequence courtesy of © Gerald de Jong (2010).

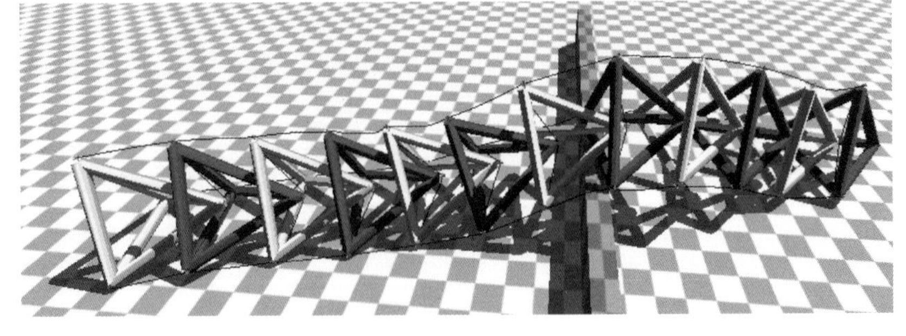

Figure 12.5
A computerized robot based on the tetrahedron and tensegrity spine shown in Figure 3.10a, page 31 Reproduced courtesy of © Tietz et al, 2013, IEEE.

Figure 12.6
A prototype of NASA's Super Ball Bot built by Ghent University's Ken Caluwaerts. Reproduced courtesy of Vytas SunSpiral and NASA Ames/ Eric James (see Figure sources and permissions).

Another robot, planned for the exploration of Titan, Saturn's largest moon, is essentially based on a collapsible tensegrity-icosahedron that unfolds during landing and adjusts its cable lengths to initiate and control movement. It does so by continually changing its position of equilibrium (SunSpiral et al., 2013) (Figure 12.6), which is particularly interesting from our perspective, because it uses the most basic of models to produce something that behaves in a very organic-like way. The complex computer algorithms necessary for controlling these robots may also provide useful information on the neuro-mechanics of human motion.

The functional human

The real significance of tensegrity models is that they simply and clearly show how every part of the body can be integrated into a complete functioning unit and they do this in remarkably life-like ways. Computerised models and the simulation of real life activities are undoubtedly the way forward but it is still the basic geometric shapes, and vector systems used to model the forces within them that help us to visualise living structures as balanced energy networks and displays of the invisible forces that hold them together (de Jong, 2011).

The unseen core

A tensegrity view of biomechanics is a model that sweeps away the man-made constraints of classical mechanics and re-establishes biology at its very core. It recognises that normal bones move around complex helical axes that are constantly changing position (Standring, 2005, p. 110) and that are guided by the particular arrangement of soft tissues surrounding them (Valero-Cuevas et al, 2009; Levin & Martin, 2012; Scarr, 2012).

A biotensegrity system enables multiple hierarchical levels to be integrated into a complete functional unit – from molecules in the cellular cytoskeleton to the entire organism – and form a compliant and flexible structure that is highly resilient to the effects of external forces. The continuous tension network is balanced by the discontinuous compression of bones and other tissue compartments, and the inherent separation of these forces means that each 'component' can be optimised to the loads that it has to carry; which makes them both light in weight and materially very efficient. A tensegrity configuration automatically balances in a position of stable equilibrium, creates an energy-efficient control strategy that is built into the structure itself and contributes to structural homeostasis (Silva et al., 2010). Why would nature *not* select such a mechanism?

Biotensegrity reveals the facts of life that we were never taught and gives new insights into the workings of the human body. The application of such knowledge is bound to have far-reaching consequences. As Paracelsus (1493–1541) said: *"the mere looking at externals is a matter for clowns, but the intuition of internals is a secret which belongs to physicians"* (Waite, A.E. ed., 1894) – and to anyone interested in biology – with the key to a deeper understanding of the mechanics and structural energetics of living organisms hiding in the geometry.

Biotensegrity: the functional basis of life

Modern anatomy has accumulated a body of knowledge that is unrivalled in any other sphere. It has classified structures according to the thinking of the day and sought to understand their functions with the latest technologies. As things have progressed, so the names and perceived functions have changed, with new aspects introduced, but these take time to be assimilated and the imagery associated with the old ones often persists. The outdated 'musculo-skeletal' duality is one such entity that has been replaced with a new name, the *mesokinetic* system, because it links bones, muscles and connective tissues into a complete functional unit; and *biotensegrity* explains the mechanics of how they can do this. The anatomy remains unchanged but our understanding of it is now different.

This book is about anatomy: the study of living form and structure; kinematics: the geometry of motion; and energetics, with biotensegrity being the overriding principle that unites them. Biological tensegrity is based on the laws of physics first, rather than the reductive methods that have dominated biomechanics for centuries; and recognises that the structure and behaviour of each molecule, cell, tissue and organism must result from those same rules.

Both simple molecules and complex structures result from the interactions of pure energy, and although particular configurations dominate, they are not especially chosen by nature but because their simplicity, efficiency and stability favours them.

An organism's survival depends on the reliability of all its parts – from molecules to the entire mesokinetic system – and is an arrangement that has refined itself over billions of years. The real beauty of nature is that it does so much with so little!

Making tensegrity models

Tensegrity models are simple (in principle) but there are a myriad different types and ways of making them, and although computers are likely to guide research into the future, it is still the stick-and-string models that really give a 'feel' for what biotensegrity is all about. Essentially the only things that are needed are compression struts, tension cables and a means of fixing them together, but, things can get a bit more complicated when building the more exotic models.

Hand-size models are fairly straight forward, particularly if using light wooden dowels and elastic cord, and the only tools needed to get started are a pair of scissors, a small saw and some cyano-acrylate glue ('superglue'); and perhaps a pair of safety glasses. The best materials that I have used so far are 3–10 mm wooden dowel (kebab sticks or a craft supplier) and 1 mm cord with a bit of stretch in it (bead and jewellery suppliers), some small elastic bands will also be needed; but essentially, use whatever is to hand.

Construction

The most useful model to make is the tensegrity-icosahedron because it easily demonstrates all the basic principles (Figure A.1). Start by selecting six struts and squaring off their ends, if necessary, and make a single cut (slot) with the saw on each end. The cord will eventually be stretched so that it just passes into these slots, and a drop of superglue is applied to finally secure them in place. Figure A.2 shows the stages of its construction.

Figure A.1
Tensegrity-icosahedron

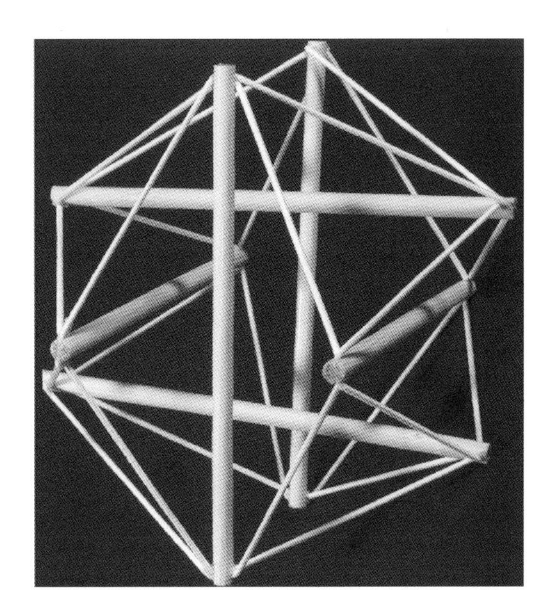

Part 1

1. First, take two struts and wrap an elastic band around one of their ends so that they are held together as a pair. Repeat with the other struts (Figure A.2a).

2. Using one pair of struts, separate the free ends and place a second pair of struts between them finally securing the ends of the first pair with another elastic band (Figure A.2b).

3. Now take the third pair of struts and separate their free ends, slipping each strut into the space on either side of the first pair (Figure A.2c), and secure all the free ends with elastic bands (Figure A.2d).

You will notice that the model now forms three sets of parallel struts with a space between each one, and all that is left is to connect them together with the cables.

Part 2

4. First, take a piece of cord and tie the two ends together so that they form a loop with each half being about 1.2 times the length of a strut (Figure A.2e). Repeat so that there are now six loops. A large elastic band would be easier but they do not last as long.

5. Stretch the cord slightly and pass it into one of the slots (nodes), then place the middle of the loop through the slots in a different pair of struts, finally connecting to the slots at the other end of the original pair (Figure A.2f).

6. Repeat step 5 (Figure A.2g) so that each slot (node) ends up with two cords i.e. two cables either side.

Figure A.2
Steps in the construction of a tensegrity-icosahedron (based on a sequence by Jon Chui (Erickson)

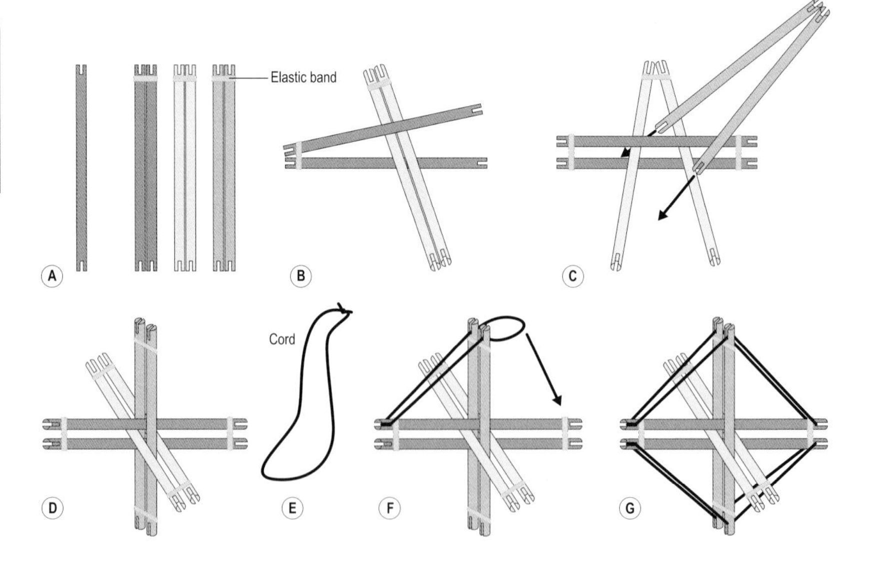

7. All that remains now is to cut the original elastic bands *carefully* with a pair of scissors and the tensegrity icosahedron will open out into the final model (Figure A.1). Finally, adjust all the cables so that they are balanced in length and place a drop of glue into each notch.

Experimenting

Daniele-Claude Martin (2012) also describes a different way to make these models. When you get really familiar with this particular arrangement of cables and struts, a single length of cord that connects each node in sequence will achieve the same result (and is overall much quicker, although fiddly to start with). Ingenuity and experimentation is the key, because if you can make a tensegrity-icosahedron, an infinite number of models become possible and will stimulate new ideas and ways of understanding biology. Have fun with them!

All photos and illustrations by Graham Scarr with the exception of the following:

Figure 1.2
OBMOKhU exhibition, 1921, Moscow. Public domain; reproduced from Gough, M. 2005. *The artist as producer: Russian constructivism in revolution*. University of California Press.

Figure 1.3
Spatial construction by Karl Ioganson 1920-1921. Public domain; copy from ViacheslavKoleichuk asreproduced in Gough, M. 2005. *The artist as producer: Russian constructivism in revolution.* University of California Press.

Figure 1.4
Superstar by Kenneth Snelson. Stephen M. Levin collection; photo © Graham Scarr.

Figure 1.8
Montreal Biosphere. Reproduced from © D.-C. Martin: Biotensegrity, KIENER, Munich 2014, with permission.

Figure 2.1
Triangulated hexagons. © Rory James – Dark-Light Photography 2013.

Figure 2.3a
Hexagonal basalts, Giant's Causeway, Northern Ireland. Reproduced from © Chmee2,http://en.wikipedia.org/wiki/Giant%27s_Causeway. Licensed under the Creative Commons Attribution 3.0 Unported (http://creativecommons.org/licenses/by/3.0/deed.en) license.

Figure 2.7
Tetrahedra. © Rory James – Dark-Light Photography 2013.

Figure 2.10
Tetrahedra with octahedra. © Rory James – Dark-Light Photography 2013.

Figure 2.11a,b
Octahedra. © Rory James – Dark-Light Photography 2013.

Figure 2.12
Octet truss. © Rory James – Dark-Light Photography 2013.

Figure 2.13a
Cube lattice. © Rory James – Dark-Light Photography 2013.

Figure 2.13b
Pyrite crystals. DonEdwards collection; photo © Graham Scarr.

Figure 2.14
Cuboctahedra. © Rory James – Dark-Light Photography 2013.

Figure 2.15
Icosahedra. © Rory James – Dark-Light Photography 2013.

Figure 2.16a
Human sapoviruses. Reproduced from © GrahamColm,http://en.wikipedia.org/wiki/Sapovirus. Licensed under the Creative Commons Attribution 3.0 Unported(http://creativecommons.org/licenses/by/3.0/deed.en) license.

Figure 2.16b
Circogonia icosahedra. Public domain; reproduced from Haeckel, E. 1887. Report on the scientific results of the voyage of H.M.S. Challenger during the years 1873-76; Zoology - volume XVIII. *Her Majesty's Stationary Office.*

Figure 2.16c
Clathrin. Reproduced from © Phoebus87,http://en.wikipedia.org/wiki/Clathrin. Licensed under the Creative Commons Attribution 3.0 Unported (http://creativecommons.org/licenses/by-sa/3.0/deed.en) license.

Figure 2.17b
Morning Glory pollen (Ipomoea purpurea). Public domain; reproduced from Dartmouth electron microscope facility, http://en.wikipedia.org/wiki/Pollen.

Figure 2.18
Jitterbug. © Rory James – Dark-Light Photography 2013.

Figure 2.19
Jitterbug (lattices). © Rory James – Dark-Light Photography 2013.

Figure 3.1
Tensegrity prisms. © Rory James – Dark-Light Photography 2013.

Figure 3.2
Tensegrity helix. © Rory James – Dark-Light Photography 2013.

Figure 3.3b
Tensegrity icosahedron. © Rory James – Dark-Light Photography 2013.

Figure 3.4
Tensegrity icosahedron with 'tension triangles'. © Rory James – Dark-Light Photography 2013.

Figure 3.5a
12-strut tensegrity sphere.© Marcelo Pars,http://www.tensegriteit.nl with permission.

Figure 3.5b
30-strut tensegrity icosahedron. © Rory James – Dark-Light Photography 2013.

Figure 3.6
Tensegrity icosahedra stretch/compression. © Rory James – Dark-Light Photography 2013.

Figure 3.7a,b
Tensegrity icosahedron hierarchies. © Rory James – Dark-Light Photography 2013.

Figure 3.8
Muscle hierarchy. Reproduced with modifications from *Journal of Bodywork and Movement Therapies* 14(4), Scarr, G., Simple geometry in complex organisms, pp. 424-44., © (2010), with permission from Elsevier.

Figure 3.10
Tensegrity models of the human spine and leg.© T. Flemons 2006, with permission. http://www.intensiondesigns.com/

Figure 4.8a
Strandbeest by Theo Jansen. © Theo Jansen, with permission.

Figure 4.8b
Strandbeest model. Theo Jansen collection;photo © Graham Scarr.

Figure 4.9a
4-bar tensegrity hierarchy. © Rory James – Dark-Light Photography 2013.

Figure 4.9b
T-icosahedron kinematics. © Rory James – Dark-Light Photography 2013.

Figure 5.1
Bovine pulmonary artery endothelial cells. Public domain; reproduced fromhttp://en.wikipedia.org/wiki/Cytoskeleton.

Figure 5.2
Tensegrity icosahedron with nucleus. © Rory James – Dark-Light Photography 2013.

Figure 5.6
Cell movement. Reproduced from *International Review of Cytology* 150, Ingber, D.E., Dike, L., Hansen, L., Karp, S., Liley, H., Maniotis, A., McNamee, H., Mooney, D., Plopper, G., Sims, J. and Wang, N. Cellular tensegrity: exploring how mechanical changes in the cytoskeleton regulate cell growth, migration and tissue pattern during morphogenesis, pp. 173-224., © (1994) with permission from Elsevier.

Figure 6.1a,c,d
Tetrahedron and T6-helix. © Rory James – Dark-Light Photography 2013.

Figure 6.5a
Geodesic forms in cytoskeleton. Reproduced with permission from © Donald E. Ingber.http://www.scholarpedia.org/article/Tensegrity.

Figure 6.7
Collagen hierarchy. Reproduced with modifications from *Journal of Bodywork and Movement Therapies* 14(4), Scarr, G., Simple geometry in complex organisms, pp. 424-44., © (2010), with permission from Elsevier.

Figure 6.9
Tube bending. © Rory James – Dark-Light Photography 2013.

Figure 6.13b
Pangolin. Reproduced from *International Journal of Osteopathic Medicine* 16, Scarr, G., Palpatory phenomena in the limbs: a proposed mechanism, pp. 114-120., © (2013), with permission from Elsevier.

Figure 7.1a
Human skeleton. Public domain; reproduced from *De Humani Corpora Fabrica*, Vesalius, A. 1543;U.S. National Library of Medicine, Bethesda, Maryland.

Figure 7.1b
Lever drawings. Public domain; reproduced from *De Motu Animalium*, Borelli, G.A, 1680; U.S. National Library of Medicine, Bethesda, Maryland.

Figure 7.6
Trampoline Club du Dauphiné: Éleonore Lachaud and Clara Guinard performing in 2007. Reproduced from ©Trampoline club Dauphine,http://en.wikipedia.org/wiki/Gymnastics.Licensed under the Creative Commons Attribution-Share Alike 3.0 Unported (http://creativecommons.org/licenses/by-sa/3.0/deed.en) license.

Figure 7.8
Arthroscopic views of the knee. Reproduced with modifications from http://www.biotensegrity.com/in_vivio_observation_of_knee_joints.php with permission from © Stephen M. Levin.

Figure 7.11
Tensegrity elbow model. Reproduced from *International Journal of Osteopathic Medicine* 15, Scarr, G., A consideration of the elbow as a tensegrity structure, pp. 53-65., © (2012), with permission from Elsevier.

Figure 8.9
Parabronchi. Reproduced from *Respiratory Physiology and Neurobiology* 155(1), Maina, J.N., Spectacularly robust! Tensegrity principle explains the mechanical strength of the avian lung, pp. 1-10., © (2007) with permission from Elsevier.

Figure 9.6
Curved-strut tensegrity hierarchy. © Rory James – Dark-Light Photography 2013.

Figure 9.7
Curved-strut tensegrity. © Rory James – Dark-Light Photography 2013.

Figure 10.14b
T-12 prism. © Rory James – Dark-Light Photography 2013.

Figure 10.14c
Tensegrity model. © Rory James – Dark-Light Photography 2013.

Figure 10.16a
Klein bottle drawing. Reproduced from © Tttrung, http://en.wikipedia.org/wiki/Klein_bottle.Licensed under the Creative Commons Attribution-Share Alike 3.0 Unported (http://creativecommons.org/licenses/by-sa/3.0/deed.en) license.

Figure 10.16b
Tensegrity Klein bottle. © Rory James – Dark-Light Photography 2013.

Figure 11.4
Leosvel and Diosmani Cuban artists and the Chinese Pole, Grenoble, 2011. Reproduced from © Ludovic Péron, http://en.wikipedia.org/wiki/Chinese_pole. Licensed under the Creative Commons Attribution 3.0 Unported(http://creativecommons.org/licenses/by-sa/3.0/deed.en) license.

Figure 12.3
Romanesque cauliflower. Reproduced from © AVM,http://en.wikipedia.org/wiki/Cauliflower. Licensed under the Creative Commons Attribution 3.0 Unported (http://creativecommons.org/licenses/by-sa/3.0/deed.en) license.

Figure 12.4
Bouncing tensegrity sphere. © Gerald de Jong 2014.

Figure 12.5
Tetraspine. © 2013 IEEE.Reprinted, with permission, from Tietz, B.R., Carnahan, R.W., Bachmann, R.J., Quinn, R.D. and SunSpiral, V. Tetraspine: robust terrain handling on a tensegrity robot using central pattern generators. *IEEE/ASME International Conference on Advanced Intelligent Mechatronics (AIM)* Wollongong, Australia.

Figure 12.6
Super Ball Bot. NASA Ames/Eric James. Research performed by Vytas SunSpiral, Adrian Agogino, and George Gorospe of NASA Ames, in the Dynamic Tensegrity Robotics Lab; Jonathan Bruce of UC Santa Cruz; Drew Sabelhaus and Alice Agogino of UC Berkeley; Atil Iscen of Oregon State University; George Korbel, Sophie Milam, Kyle Morse, and David Atkinson of the University of Idaho; and the model built by Ken Caluwaerts of Ghent University.

Figure A.1
Tensegrity icosahedron. © Rory James – Dark-Light Photography 2013.

Abduction
the action of moving an arm or leg outwards from the side

Actin
a globular protein (g-actin) that can polymerise into long fibres (f-actin) called microfilaments

Actin bundle
a group of actin microfilaments laid side by side that are able to withstand compression

Actomyosin motors
a combination of the proteins actin and myosin that adjust tension within the cytoskeleton

Adhesion molecule
a transmembrane protein that connects the internal cytoskeleton to the extracellular matrix and adjacent cells

Agonist
a muscle that contracts in combination with others to cause a particular movement

Alpha helix
a specific type of helical winding where a polypeptide chain is stabilised by hydrogen bonding

Alveolus
a terminal air cell in the mammalian lung that is separated from the surrounding capillaries by a single cell and basement membrane

Amino acid
one of the twenty-four 'building blocks' of proteins

Amman lines
the arrangement of lines that cross the surface of Penrose tilings are separated by a spacing that is related to the Fibonacci sequence and confirms that the pattern is quasi-periodic

Anastomosis
an interconnected network of blood vessels or nerves

Anconeus
an elbow muscle considered to play a role in maintaining the spacing between the humerus and ulna

Ankyrin
a family of adaptor proteins that mediate the attachment of spectrin to the underlying cell membrane

Antagonist
a muscle whose contraction causes movement in the 'opposite' direction to another muscle

Antiprism
a polyhedron consisting of two similar and parallel polygons connected by triangular faces

Apical dome
the bulging part of a cell immediately behind the formation of a lamellipodium

Aponeurosis
a sheet-like fascial expansion connecting a muscle to other anatomical structures

Apoptosis
programmed cell death

Articulation
the moveable connections between two bones

Basement membrane
a specialised type of extra-cellular matrix lying beneath epithelial cells

Buckling
the result of bending or twisting a material out of shape

Buckyball
a molecule consisting of sixty carbon atoms arranged as a truncated icosahedron

Cadherins
one of several different connectors that link adjacent cells

Capsule
the fibrous connective tissue that encloses synovial joints

Cell cortex
the part of the cytoskeleton lying immediately beneath the cell membrane

Cephalopod
a class of bisymmetrical molluscs, e.g. nautilus and squid

Cerebellum
a part of the hindbrain in vertebrates involved in the control of movement

Chirality
a left or right-handed twist

Chondrocyte
a cartilage cell

Coiled-coil
a second-order helical winding

Covalent bonding
a chemical bond whereby two atoms are held together by interactions of their nuclei and surrounding electrons
Dimer
a particular molecular species formed by two subunits
Distal
an indication of a particular region in a limb that is furthest away from the body
Extra-cellular matrix
the fibrous matrix that surrounds cells and includes the basement membrane and interstitial matrix; it is a defining feature of connective tissues
Ectomeninx
the layer of mesodermal tissue surrounding the embryonic vertebral brain and from which membranous bone, dura mater and periosteum are derived
Elastin
a particular protein in many connective tissues that allows a tissue to return to its usual shape after distortion
Embryogenesis
the process leading to the formation of the embryo
Endomysium
the intramuscular connective tissue that surrounds and unites individual muscle fibres
Endoskeleton
the internal skeletal support system e.g. cartilage and bone in vertebrates
Enzyme
a catalytic protein that increases the activity of a particular chemical reaction
Epimysium
the tough fascial connective tissue surrounding muscle and that is continuous with the intra-muscular perimysium and surrounding inter-mysial and extra-mysial fascia
Epithelial-mesenchymal-transition
the process by which epithelial cells lose their polarity and adhesion to other cells, and gain migratory and invasive properties to become mesenchymal cells
Epithelium
a form of tissue characterised by an arrangement of cells fixed to an underlying basement membrane; it tends to line the inner surfaces of cavities
Eukaryote
a higher organism with cells containing a nucleus bounded by a membranous nuclear envelope and with many cytoplasmic organelles; in contrast to prokaryotes such as bacteria and blue-green algae
Exoskeleton
an external skeletal support system e.g. chitin in insects
Extensor
a muscle that increases the angle of a joint towards 180°
Fascicle
a bundle of fibres
Femoro-meniscal
the articulation between the femur bone and meniscal cartilages
Femur
the thigh bone
Fibril
a small fibre
Fibrillogenesis
the formation of a structure made from fibrils
Fibronectin
a complex protein within the extracellular matrix that links collagen fibres with integrins in the cell membrane
Filopodia
a filamentous spike originating from the surface of some cells
Flexor
a muscle that tends to reduce the angle between two bones to 0 degrees
Fractal
a mathematical construct with dimensions between 1 and 2, or 2 and 3, etc. that is often compared to the self-similar patterns that appear at different size scales in biological structures
Free-body analysis
a geometric analysis that simplifies a complex of force vectors to a manageable minimum
Frequency
a representation of expanding energy levels as used by Buckminster Fuller to understand the relationship between simplicity and complexity

Fullerene
a molecule with carbon atoms configured as a geodesic dome, as described by Buckminster Fuller
Gleno-humeral
the joint between the scapula (shoulder blade) and humerus (arm bone)
Glenoid
the part of the scapula that articulates with the humerus
Glycine
the smallest of all amino acids and the only one that does not have a chiral counterpart
Glycolysis
the sequence of reactions involved in the breakdown of glucose to form energy
Humeral head
the top of the arm bone that forms part of the shoulder
Humerus
the arm bone
Hydrophilic
the attraction to water
Ischium
one of the bilateral 'sitting' bones that form part of the pelvis
Iliac crest
the top part of the bilateral iliac bones that form part of the pelvis
Infundibulum
the funnel-shaped opening between the atrium and exchange tissues in the avian lung
Lamellae
thin layers having differing characteristics
Lamellipodia
a protrusion that extends outwards from a cell
Liquid crystal
a liquid or soft-tissue that forms a regular structure under certain conditions (quasi-crystal)
Mendelian rules
a system of rules discovered by Gregor Mendel that determine the inheritable characteristics of an organism
Meniscus
a relatively loose cartilage sheet found in the knee, etc.
Messenger RNA
a form of ribonucleic acid that transfers information from DNA to be transcribed into proteins
Metaphysical
a conceptual reality that is indirectly linked to the physical realm
Micelles
an aggregate of molecules that collect in a roughly spherical form
Mitogen
a chemical agent that stimulates a change in a particular response within a cell
Möebius strip
a one-side surface formed by twisting one end of a strip and joining it to the other end
Myofibroblast
a mesenchymal cell that generates tension within the extracellular matrix
Nanofibril
a small fibre at the nano-scale (10^{-6} metres)
Nano-structure
a structure at the nano-scale (10^{-6} metres)
Notochord
embryonic precursor to the vertebral column
Omni-directional
occurring in every direction around a sphere
Osteoblast
a mesenchymal cell involved in the formation of bone
Patella
the knee cap, a sesamoid bone found at the front of the knee
Patello-femoral
the region between the femur and knee cap
Penrose tiling
a form of tiling discovered by Roger Penrose with 5-fold symmetry that never repeats itself exactly in any direction
Pentamer
a molecule consisting of five subunits
Perichondrial cells
the cells surrounding the developing cartilage
Perimysium
the fibrous connective tissue that encloses bundles or fasciculi of muscle fibres, and transmits the force of muscle fibre contraction to the outer epimysium and tendon
Periosteum
the tough layer of collagen fibres covering the surface of bone and is very pain sensitive

Polymerisation
the connection of similar size particles into a long filament or tubule

Pre-stress
the intrinsic tension found within living tissues

Proprioception
the sensory apparatus that detects changes in force, shape and position

Proteoglycan
a complex molecule that associates with collagens in the extracellular matrix

Proximal
an indication of a part of a limb that is closest to the body

Pubis
the anterior junction between the two pubic bones at the front of the pelvis

Quadriceps
the four muscles covering the front of the thigh

Sacro-ilium
the region on either side of the sacrum next to the ilia bones of the pelvis

Sacro-spinous ligaments
the tough connective tissues between the sacrum and iliac bones

Sacro-tuberous ligaments
the tough connective tissues between the sacrum and ischial tuberosities

Scapula
shoulder blade

Serration
like a saw tooth

Sesamoid bone
bones that are embedded in tendons or other dense connective tissues

Shear stress
the resultant of two force vectors that run alongside each other but in opposite directions

Signalling cascade
a change in activity in one molecule that stimulates a reaction in a distant one through a series of interlinked molecules

Stem cell
an undifferentiated cell that can give rise to a variety of different cell species

Sterno-clavicular joint
the region between the sternum (breastbone) and clavicle (collar bone)

Steric repulsion
the effect of a molecule's size in maintaining a certain distance from other molecules

Synergist
the co-operative effect of two muscles working together to create a different outcome than that of each one on its own

Synovial joint
a region between two bones surrounded by a fibrous capsule containing synovial fluid that permits changes in position relative to each other

Tetramer
a molecule that consists of four connected parts

Thoraco-lumbar
the region in the back between the thorax (ribcage) and lumbar vertebrae

Tibia
the largest bone below the knee

Transcription
the conversion of one particular molecular pattern into a different one, e.g. the DNA code is transcribed into a messenger RNA code for the eventual production of a protein

Transmembrane
a molecule that has active sites on both sides of the cell membrane

Transverse
a section that is horizontal or perpendicular to the long axis

Trimer
a molecule that consists of three subunits

Trochlear notch
the concave depression in the ulna part of the elbow joint

Ulna
one of the bones in the forearm

Van der Waal
the weak (but significant) nuclear forces between crystals and molecules

Vertex
the point of intersection of the sides and faces of a polygon

1. Alfaro, M.E., Bolnick, D.I. and Wainwright, P.C. (2004) Evolutionary dynamics of complex biomechanical systems: an example using the four-bar mechanism. *Evolution* 58(3), pp. 495–503.
2. Alippi, A., Biagioni, A., Conclusio, D. and D'Orazio, A. (2010) Nonlinear phenomena in vibrating tensegrity structures. *39th International Congress on Noise Control Engineering 2010: (Internoise, 2010)*. Lisbon, Portugal, 13–15 June. 9, pp. 6831–6836.
3. Andrews, S., Shrive, N. and Ronsky, J. (2011) The shocking truth about meniscus. *Journal of Biomechanics* 44(16), pp. 2737–2740.
4. Atela, P., Golé, C. and Hotton, S. (2002) A dynamical system for plant pattern formation: a rigorous analysis. *Journal of Nonlinear Science* 12, pp. 641–676.
5. Bassnett, S., Missey, H. and Vucemilo, I. (1999) Molecular architecture of the lens fiber cell basal membrane complex. *Journal of Cell Science* 112(13), pp. 2155–2165.
6. Baudriller, H., Maurin, B., Cañadas, P., Montcourrier, P. and Parmeggiani, A. (2006) Form-finding of complex tensegrity structures: application to cell cytoskeleton modelling. *Comptes Rendus Mecanique* 334, pp. 662–668.
7. Benetazzo, L., Bizzego, A., De Caro, R., Frigo, G., Guidolin, D. and Stecco, C. (2011) 3D reconstruction of the crural and thoracolumbar fasciae. *Surgical and Radiologic Anatomy* [Online] 33(10), pp. 855–862. Available at: http://link.springer.com/article/10.1007%2Fs00276-010-0757-7#page-1 [Accessed: 8th March 2014].
8. Biewener, A.A. (1998) Muscle-tendon stresses and elastic energy storage during locomotion in the horse. *Comparative Biochemistry and Physiology Part B* 120(1), pp. 73–87.
9. Blechschmidt, E. (2004) *The ontogenetic basis of human anatomy: a biodynamic approach to development from conception to birth*. North Atlantic Books.
10. Bowick, M.J. and Giomi, L. (2009) Two-dimensional matter: order, curvature and defects. *Advances in Physics* [Online] 58(5), pp. 449–563. http://arxiv.org/abs/0812.3064 [Accessed: 8th March 2014].
11. Brangwynne, C.P., MacKintosh, K.C., Kumar, S., Geisse, N.A., Talbot, J., Mahadevan, L., Parker, K.K., Ingber, D.E. and Weitz, D.A. (2006) Microtubules can bear enhanced compressive loads in living cells because of lateral reinforcement. *Journal of Cell Biology* 173(5), pp. 733–741.
12. Buckberg, G.D. (2002) Basic science review: the helix and the heart. *Journal of Thoracic and Cardiovascular Surgery* 124(5), pp. 863–883.
13. Butcher, J.A. and Lamb, G.W. (1984) The relationship between domes and foams: application of geodesic mathematics to micelles. *Journal of the American Chemistry Society* 106, pp. 1217–1220.
14. Carey, E.J. (1920) Studies in the dynamics of histogenesis: I. Tension of differential growth as a stimulus to myogenesis. *Journal of General Physiology* 2(4), pp. 357–372.
15. Carey, E.J. (1920) Studies in the dynamics of histogenesis: II. Tension of differential growth as a stimulus to myogenesis in the esophagus. *Journal of General Physiology* 3(1), pp. 61–83.
16. Caspar, D.L.D. (1980) Movement and self-control in protein assemblies: quasi-equivalence revisited. *Biophysics Journal* 32(1), pp. 103–133.
17. Castle, T., Evans, M.E., Hyde, S.T., Ramsden, S. and Robins, V. (2012) Trading spaces: building three-dimensional nets from two-dimensional tilings. *Journal of the Royal Society Interface Focus* 2(2), pp. 555–566.
18. Cavalcante, F.S.A., Ito, S., Brewer, K., Sakai, H., Alencar, A.M., Almeida, M.P., Andrade, J.S., Majumdar, A., Ingenito, E.P. and Suki, B. (2005) Mechanical interactions between collagen and proteoglycans: implications for the stability of lung tissue. *Journal of Applied Physiology* 98(2), pp. 672–679.
19. Charvolin, J. and Sadoc, J.F. (2011) A phyllotactic approach to the structure of collagen fibrils. *Biophysical Review Letters* 6(01n02), pp. 13–27.
20. Chen, C.S. and Ingber, D.E. (1999) Tensegrity and mechanoregulation: from skeleton to cytoskeleton. *Osteoarthritis and Cartilage* 7(1), pp. 81–94.
21. Chen, T.J., Wu, C.C., Tang, M.J., Huang, J.S. and Su, F.C. (2010) Complexity of the tensegrity structure for dynamic energy and force distribution of cytoskeleton during cell spreading. *PloS ONE* [Online] 5(12), e14392. Available at: http://www.plosone.org/article/fetchObject.action?uri=info%3Adoi%2F10.1371%2Fjournal.pone.0014392&representation=PDF [Accessed: 8th March 2014].
22. Chouaieb, N., Goriely, A. and Maddocks, J.H. (2006) Helices. *Proceedings of the National Academy of Sciences* 103(25), pp. 9398–9403.
23. Clark, R.B. and Cowey, J.B. (1958) Factors controlling the change of shape of certain nemertean and turbellarian worms. *Journal of Experimental Biology* 35(Dec.), pp. 731–748.
24. Connelly, R. and Back, A. (1998) Mathematics and tensegrity. *American Scientist* 86(2), pp. 42–151.
25. Cretu, S.M. (2009) Tensegrity as a structural framework in life sciences and bioengineering. In: Awrejcewicj J. ed. *Modeling, simulation and control of nonlinear engineering dynamical systems* Springer, pp. 301–311.
26. Cretu, S.M. and Catalina, B.G. (2011) Tensegrity applied to modelling the motion of viruses. *Acta Mechanica Sinica* 27(1), pp. 125–129.
27. Crick, F.H.C. and Watson, J.D. (1956) Structure of small viruses. *Nature* 177(10th March), pp. 473–475.
28. Cummings, C.H. (1994) A tensegrity model for osteopathy in the cranial field. *Journal of the American Association of Osteopathy* (Spring), pp. 9–27.
29. Darwin, C. (1858) *On the origin of species by means of natural selection.* London, John Murray.
30. Davies, E. (2004) Tensegrity: architecture to biomechanics. *The Osteopath* (Oct.), pp. 29–31.
31. Davies, E. (2004) Tensegrity 2: A new model of the structure function relationship. *The Osteopath* (Dec.), pp.41–43.
32. De Jong, G. (2010) Tensegrity sphere falling bouncing. [Online] Available at: http://www.youtube.com/watch?v=-6l3utbJ1M8 [Accessed 8th March 2014].
33. De Jong, G. (2011) Darwin at home. [Online] Available at: https://www.youtube.com/watch?v=_Il-uESToOs [Accessed 21st March 2014].
34. Denton, M.J., Marshall, C.J. and Legge, M. (2002) The protein folds as Platonic forms: new support for the pre-Darwinian conception of evolution by natural law. *Journal of Theoretical Biology* 219, pp. 325–342.
35. Denton, M.J., Dearden, P.K. and Sowerby, S.J. (2003) Physical law *not natural selection* as the major determinant of biological complexity in the subcellular realm: new support for the pre-Darwinian conception of evolution by natural law. *BioSystems* 71, pp. 297–303.
36. De Oliveira, M., Vera, C., Valdez, P., Sharma, Y., Skelton, R. and Sung, L.A. (2010) Nanomechanics of multiple units in the erythrocyte membrane skeletal network. *Annals of Biomedical Engineering* 38(9), pp. 2956–2967.

37. De Santis, G., Lennon. A.B., Verhegghe, B., Verdonck, P. and Prendergast, P.J. (2011) How can cells sense the elasticity of a substrate? An analysis using a cell tensegrity model. *European Cells and Materials Journal* 22(Oct), pp. 202–213.
38. De Varco, B.G. (date unknown) Invisible architecture: the nanoworld of Buckminster Fuller. [Online] Available at: http://members.cruzio.com/~devarco/invisible.htm [Accessed: 8th March 2014].
39. Douady, S. and Couder, Y. (1996) Phyllotaxis as a dynamical self-organizing process; part I: the spiral modes resulting from time-periodic iterations. *Journal of Theoretical Biology* 178, pp. 255–274.
40. Du, N., Liu, X.Y., Narayanan, J., Li, L., Lim, M.L.M. and Li, D. (2006) Design of superior spider silk: from nanostructure to mechanical properties. *Biophysical Journal* 91, pp. 4528–4535.
41. DuFort, C., Maszek, M.J. and Weaver, V.M. (2011) Balancing forces: architectural control of mechanotransduction. *Nature Reviews Molecular Cell Biology* 12(5), pp. 308–319.
42. Eckstein, F., Löhe, F., Hillebrand, S., Bergmann, M., Schulte, E., Milz, S. and Putz, R. (1995a) Morphomechanics of the humero-ulnar joint: I. Joint space width and contact areas as a function of load and flexion angle. *The Anatomical Record* 243(3), pp. 318–326.
43. Eckstein, F., Merz, B., Müller-Gerbl, M., Holzknecht, N., Pleier, M. and Putz, R. (1995b) Morphomechanics of the humero-ulnar joint: II. Concave incongruity determines the distribution of load and subchondral mineralization. *The Anatomical Record* 243(3), pp. 227–335.
44. Edmondson, A.C. (2007) *A Fuller explanation: the synergetic geometry of R. Buckminster Fuller.* Colorado: Emergent World Press.
45. Edwards, S.A., Wagner, J. and Gräter, F. (2012) Dynamic prestress in a globular protein. *PloS ONE* [Online] 8(5), e1002509. Available at: http://www.ploscompbiol.org/article/fetchObject.action?uri=info%3Adoi%2F10.1371%2Fjournal.pcbi.1002509&representation=PDF [Accessed on: 8th March 2014].
46. Emons, A.M. and Mulder, B.M. (1998) The making of the architecture of the plant cell wall: how cells exploit geometry. *Proceedings of the National Academy of Sciences* 95, pp. 7215–7219.
47. Erickson, A. (Online) Available at: www.geoburst.ca [Accessed 16th March 2014].
48. Ernst, B. (1985) *The magic mirror of M.C.Escher.* Norfolk, Tarquin publications.
49. Ethier, C.R. and Simmons, C.A. (2007) *Introductory Biomechanics: from cells to organisms.* Cambridge University Press.
50. Findley, T.W. and Shalwala, M. (2013) Fascia research congress evidence from the 100 year perspective of Andrew Taylor Still. *Journal of Bodywork and Movement Therapies* 17(3), pp. 356–364.
51. Fischer, S., Exner, A., Zielske, K., Perlich, J., Deloudi, S., Steurer, W., Lindner, P. and Förster, S. (2011) Colloidal quasicrystals with 12-fold and 18-fold diffraction symmetry. *Proceedings of the National Academy of Sciences* 108, pp. 1810–1814.
52. Fabre, A. et al. (2013) Influence of body mass on the shape of forelimb in musteloid carnivorans. *Biological Journal of the Linnean Society* 110, pp. 91–103.
53. Flemons, T. (2007) The geometry of tensegrity. [Online] Available at: http://www.intensiondesigns.com/geometry_of_anatomy.html [Accessed 24th March 2014].
54. Flemons, T. (2012) The bones of tensegrity. [Online] Available: http://www.intensiondesigns.com/bones_of_tensegrity.html [Accessed 8th March 2014].
55. Fotin, A., Kirchhausen, T., Grigorieff, N., Harrison, S.C., Walz, T. and Cheng, Y. (2006) Structure determination of clathrin coats to subnanometer resolution by single particle cryo-electron microscopy. *Journal of Structural Biology* 156(3), pp. 453–460.
56. Fratzl, P. ed. (2008) *Collagen: structure and mechanics*. Springer.
57. Fuller, R.B. (1975) *Synergetics, Explorations in the Geometry of Thinking.* Macmillan.
58. Fuller, R.B. (1976) "Synergetics." Exhibition catalog, Cooper Hewitt Museum of Design.[Online] Available at: http://bfi.org/about-fuller/big-ideas/synergetics/tensegrity-geometry-thinking [Accessed 19th March 2014].
59. Gabella, G. (1987) The cross-ply arrangement of collagen fibres in the submucosa of the mammalian small intestine. *Cell and Tissue Research* 248(3), pp. 491–497.
60. Galilei, G. (1638) *Discourses and mathematical demonstrations relating to two new sciences.* Louis Elsevir.
61. Gao, H., Ji, B., Jäger, I.L., Arzt, E. and Fratzl, P. (2003) Materials become insensitive to flaws at nanoscale: lessons from nature. *Proceedings of the National Academy of Sciences* 100(10), pp. 5597–5600.
62. Garcia-Ellis, M., Lluch, A.L., Ferreres, A., Lluch, F.W. and Lhamby, F. (2008) Transverse loaded pronosupination test. *Journal of Hand Surgery* (European Volume) 33E(6), pp. 765–767.
63. Gibson, C.M., Gibson, W.J., Murphy, S.A., Marble, S.J., McCabe, C.H., Trakhia, M.P., Kirtane, A.J., Karha, J., Aroesty, J.M., Giugliano, R.P. and Antman, E.M. (2003) Association of the Fibonacci cascade with the distribution of coronary artery lesions responsible for ST-segment elevation myocardial infarction. *American Journal of Cardiology* 92(Sept), pp. 595–597.
64. Gleik, J. (1997) *Chaos: making a new science.* Vintage.
65. Gordon, J.E. (1978) *Structures or why things don't fall down.* Penguin.
66. Gough, M. (2005) *The artist as producer: Russian constructivism in revolution.* University of California Press.
67. Gracovetsky, S. (1994) On the validity of mathematical modelling of the spine. Letters to the editor. *Clinical Biomechanics* 9(5), pp. 325–326.
68. Guimberteau, J.C. (2005) The sliding mechanics of the subcutaneous structures in man; illustration of a functional unit: the microvacuoles. *Studies of the Académie Nationale de Chirurgie* 4(4), pp. 35–42.
69. Gupta, H.S., Seto, J., Wagermaier, W., Zaslansky, P., Boesecke, P. and Fratzl, P. (2006) Cooperative deformation of mineral and collagen in bone at the nanoscale. *Proceedings of the National Academy of Sciences* 103(47), pp. 17741–17746.
70. Haeckel, E. (1887) Report on the scientific results of the voyage of H.M.S. Challenger during the years 1873–76; Zoology – volume XVIII. *Her Majesty's Stationary Office.*
71. Hameroff, S. and Tuszynski, J. (2004) Quantum states in proteins and protein assemblies: the essence of life? In: Abbott, D. et al., eds. *Fluctuations and Noise in Biological, Biophysical, and Biomedical Systems II* Proceedings of the International Society for optics and photonics (SPIE), 5467, pp. 27–41.
72. Hansen, L., de Zee, M., Rasmussen, J., Anderson, T.B., Wong, C. and Simonsen, E.B. (2006) Anatomy and biomechanics of the back muscles in the lumbar spine with reference to

biomechanical modeling. *Spine* 31(17), pp. 1888–1899.

73. Harris, A.K., Wild, P. and Stopak, D. (1980) Silicone rubber substrata: a new wrinkle in the study of cell locomotion. *Science* 208(April), pp. 177–179.
74. Heartney, E. (2009) *Kenneth Snelson: forces made visible*. Massachusetts, Hard Press Editions.
75. Hebrank, M.R. (1980) Mechanical properties and locomotor functions of eel skin. *Biological Bulletin* 158, pp. 58–68.
76. Hedrick, T.L., Tobalske, B.W., Ros, I.G., Warrick, D.R. and Biewener, A.A. (2012) Morphological and kinematic basis of the hummingbird flight stroke: scaling of flight muscle transmission ratio. *Proceedings of the Royal Society B* [Online] 279, pp. 1986–1992. Available: http://rspb.royalsocietypublishing.org/content/279/1735/1986.full.pdf+html [Accessed on 8th March 2014].
77. Henderson, J.H. and Carter, D.R. (2002) Mechanical induction in limb morphogenesis: the role of growth-generated strains and pressures. *Bone* 31(6), pp. 645–653.
78. Hickman, J. and Humphrey, M. (1988) *Hickman's farriery*. London, J. A. Allen.
79. Hohenschurz-Schmidt, D. (2013) Personal communication.
80. Holzapfel, G.A. (2008) Collagen in arterial walls: biomechanical aspects. In: Fratzl, P. ed. *Collagen: structure and mechanics*. Springer, pp. 285–324.
81. Huang, S. and Ingber, D.E. (1999) The structural and mechanical complexity of cell-growth control. *Nature Cell Biology* 1(Sept), pp. E131–E138.
82. Huijing, P.A. and Baan, G.C. (2003) Myofascial force transmission: muscle relative position and length determine agonist and synergist muscle force. *Journal of Applied Physiology* 94, pp. 1092–1107.
83. Huijing, P.A. and Baan, G.C. (2008) Myofascial force transmission via extramuscular pathways occurs between antagonistic muscles. *Cells Tissues Organs* 188, pp. 400–414.
84. Hukins, D.W.L. and Meakin, J.R. (2000) Relationship between structure and mechanical function of the tissues of the intervertebral joint. *American Zoologist* 40, pp. 42–52.
85. Ingber, D.E. et al. 1981. Role of basal lamina in neoplastic disorganization of tissue architecture. *Proceedings of the National Academy of Sciences* 78(6), pp. 3901–3905.
86. Ingber, D.E. (1993) Cellular tensegrity: defining new rules of biological design that govern the cytoskeleton. *Journal of Cell Science* 104(3), pp. 613–627.
87. Ingber, D.E., Dike, L., Hansen, L., Karp, S., Liley, H., Maniotis, A., McNamee, H., Mooney, D., Plopper, G., Sims, J. and Wang, N. (1994) Cellular tensegrity: exploring how mechanical changes in the cytoskeleton regulate cell growth, migration and tissue pattern during morphogenesis. *International Review of Cytology* 150, pp. 173–224.
88. Ingber, D.E. (1998) The architecture of life. *Scientific American* (Jan), pp. 30–39.
89. Ingber, D.E. (2003a) Tensegrity I. Cell structure and hierarchical systems biology. *Journal of Cell Science* 116(7), pp. 1157–1173.
90. Ingber, D.E. (2003b) Tensegrity II. How structural networks influence cellular information processing networks. *Journal of Cell Science* 116(8), pp. 1397–1408.
91. Ingber, D.E. (2003c) Mechanobiology and diseases of mechanotransduction. *Annals of Medicine* 35, pp. 1–14.
92. Ingber, D.E. (2006) Mechanical control of tissue morphogenesis during embryological development. *International Journal of Developmental Biology* 50(2–3), pp. 255–266.
93. Ingber, D.E. (2008) Tensegrity-based mechanosensing from macro to micro. *Progress in Biophysics and Molecular Biology* 97(2–3), pp. 163–179.
94. Ingber, D.E. and Folkman, J. (1989) Mechanochemical switching between growth and differentiation during fibroblast growth factor-stimulated angiogenesis in vitro: role of extracellular matrix. *Journal of Cell Biology* 109(1), pp. 317–330.
95. Isenberg, C. (1992) *The science of soap films and soap bubbles*. Dover Publications.
96. Izard, T., Ævarsson, A., Allen, M.D., Westphal, A.H., Perham, R.N., de Kok, A. and Hol, W.G.J. (1999) Principles of quasi-equivalence and Euclidean geometry govern the assembly of cubic and dodecahedral cores of pyruvate dehydrogenase complexes. *Proceedings of the National Academy of Sciences* 96, pp. 1240–1245.
97. Jamali, Y., Azimi, M. and Mofrad, M.R.K. (2010) A sub-cellular viscoelastic model for cell population mechanics. *PloS ONE* [Online] 5(8), e12087. [Available at: http://www.plosone.org/article/fetchObject.action?uri=info%3Adoi%2F10.1371%2Fjournal.pone.0012097&representation=PDF [Accessed 8th March 2014].
98. Jasinoski, S.C., Reddy, B.D., Louw, K.K. and Chinsamy, A. (2010) Mechanics of cranial sutures using the finite element method. *Journal of Biomechanics* 43(16), pp. 3104–3111.
99. Jauregui, V.G. (2010) *Tensegrity structures and their application to architecture*. Ediciones de la Universidad de Cantabria.
100. Jelinek, H.F., Jones, C.L., Warfel, M.D., Lucas, C., Depardieu, C. and Aurel, G. (2006) Understanding fractal analysis? The case of fractal linguistics. *Complexus* [Online] 3, pp. 66–73. [Available at: http://www.karger.com/Article/Pdf/94189 [Accessed 8th March 2014].
101. Johnston, J.C., Kastelowitz, N. and Molinero, V. (2010) Liquid to quasicrystals transition in bilayer water. *Journal of Chemical Physics* [Online] 133, pp. 154516. Available at: http://scitation.aip.org/content/aip/journal/jcp/133/15/10.1063/1.3499323 [Accessed 9th March 2014.
102. Jones, L.J.F., Carballido-López, R. and Errington, J. (2001) Control of *cell* shape in bacteria: helical, actin-like filaments in Bacillus subtilis. Cell 104, pp. 913–922.
103. Juan, S.H. and Tur, J.M.M. (2008) Tensegrity frameworks: static analysis review. *Mechanism and Machine Theory* 43, pp. 859–81.
104. Kadler, K.E., Hill, A. and Canty-Laird, G. (2008) Collagen fibrillogenesis: fibronectin, integrins, and minor collagens as organizers and nucleators *Current Opinion in Cell Biology* 20(5–24), pp. 495–501.
105. Kapandji, A.I. (2012) Le système conjonctif, grand unificateur de l'organisme. *Annales de Chirurgie Plastique Esthetique* 57, pp. 507–514. [French]
106. Kardon, G., Harfe, B.D. and Tabin, C.J. (2003) A Tcf4-positive mesodermal population provides a prepattern for vertebrate limb muscle patterning. *Developmental Cell* 5, pp. 937-944.
107. Kardong, K.V. (2003) Biomechanics and evolutionary space: a case study. In: Bels, V.L. et al. Eds. *Vertebrate biomechanics and evolution* Oxford, BIOS Scientific Publishers, pp.73–86.
108. Kassolik, K., Andrzejewski, W., Brzozowski, M., Wilk, I., Górecka-Midura, L., Ostrowska, B., Krzyżanowski, D. and Kurpas, D. (2013) Comparison of massage based on the tensegrity principle and classic massage in treating chronic shoulder pain. *Journal of Manipulative and*

Physiological Therapeutics 36(7), pp. 418–427.

109. Kenner, H. (2003) *Geodesic math and how to use it.* University of California.
110. Kier, W.M. and Smith, K.K. (1985) Tongues, tentacles and trunks: the biomechanics of movement in muscular-hydrostats. *Zoological Journal of the Linnean Society* 83, pp. 307–324.
111. Kim, P.T., Isogai, S., Murakami, G., Wada, T., Aoki, M., Yamashita, T. and Ishii, S. (2002) The lateral collateral ligament complex and related muscles act as a dynamic stabilizer as well as a static supporting structure at the elbow joint: an anatomical and experimental study. *Okajimas Folia Anatomica Japonica* 79, pp. 55–62.
112. Knight, D.P. and Vollrath, F. (2002) Biological liquid crystal elastomers. *Philosophical Transactions of the Royal Society of London B* 357, pp. 155–163.
113. Koehl, M.A.R., Quillin, K.J. and Pell, C.A. (2000) Mechanical design of fiber-wound hydraulic skeletons: the stiffening and straightening of embryonic notochords. *American Zoologist* 40, pp. 28–41.
114. Kovacs. F., Tarnai, T., Fowler, P.W. and Guest, S.D. (2004) A class of expandable polyhedral structures. *International Journal of Solids and Structures* 41(3–4), pp. 1119–1137.
115. Kroto, H. (1988) Space, stars, C60, and soot. *Science* 242, pp. 1139–1145.
116. Ladoux, B. and Nicolas, A. (2012) Physically based principles of cell adhesion mechanosensitivity in tissues. *Reports on Progress in Physics* 75(11), pp. 116601.
117. Lakes, R. (1993) Materials with structural hierarchy. *Nature* 361(Feb), pp. 511–515.
118. Langevin, H.M. 2006. Connective tissue: a body-wide signalling network? *Medical Hypotheses* 66(6), pp. 1074–1077.
119. Le Bozec, S., Maton, B. and Cnockaert, J.C. (1980) The synergy of elbow extensor muscles during static work in man. *European Journal of Applied Physiology* 43, pp. 57–68.
120. Levin, S.M. (1981) The icosahedron as a biologic support system. *Proceedings of the 34th Annual Conference on Engineering in Medicine and Biology* Houston, Texas, 23, p. 404.
121. Levin, S.M. (1982) Continuous tension, discontinuous compression: a model for biomechanical support of the body. *Bulletin of Structural Integration* 8.
122. Levin, S.M. (1986) The icosahedron as the three-dimensional finite element in biomechanical support. *Proceedings of the Society of General Systems Research on Mental Images, Values and Reality* University of Philadelphia, G14–G26.
123. Levin, S.M. (1995) The importance of soft tissues for structural support of the body. *Spine: State of the Art Reviews* 9(2), pp. 357–363.
124. Levin, S.M. (1997) Putting the shoulder to the wheel: a new biomechanical model for the shoulder girdle. *Journal of Biomedical and Scientific Instrumentation* 33, pp. 412–417.
125. Levin, S.M. (2002) The tensegrity truss as a model for spine mechanics: biotensegrity. *Journal of Mechanics and Medicine in Biology* 2, pp. 375–88.
126. Levin, S.M. (2005) In vivo observation of articular surface contact in knee joints. Unpublished. [Available at: www.biotensegrity.com [Accessed 8th March 2014].
127. Levin, S.M. (2006) Tensegrity: the new biomechanics. In: Hutson, M. and Ellis, R. eds *Textbook of musculoskeletal medicine*. Oxford University Press.
128. Levin, S.M. (2007) A suspensory system for the sacrum in pelvic mechanics: biotensegrity. In: Vleeming, A. et al. eds. *Movement, Stability and Lumbopelvic Pain.* 2nd edn. Edinburgh, Elsevier.
129. Levin, S.M. (2009) Letter to the editor commenting on: Masi, A.T. and Hannon, J.C. Human resting muscle tone (HRMT): narrative introduction and modern concepts. *Journal of Bodywork and Movement Therapies* 13, pp. 117–120.
130. Levin, S.M. (2014) Personal communication. [See acknowledgements].
131. Levin, S.M. and Martin, D.C. (2012) Biotensegrity: the mechanics of fascia. In: Schleip, R. eds. *Fascia: the tensional network of the human body.* Edinburgh, Elsevier, pp. 137–141.
132. Levin, S.M. and Scarr, G. (2012) unpublished.
133. Li, J., Dao, M., Lim, C.T. and Suresh, S. (2005) Spectrin-level modelling of the cytoskeleton and optical tweezers stretching of the erythrocyte. *Biophysical Journal* 88(5), pp. 3701–3719.
134. Liedl, T., Högberg, B., Tytell, J., Ingber, D.E. and Shih, W.M. (2010) Self-assembly of three-dimensional prestressed tensegrity structures from DNA. *Nature Nanotechnology* 5(7), pp. 520–524.
135. Lighthill, J. (1986) The recently recognized failure of predictability in Newtonian dynamics. *Proceedings of the Royal Society of London A* 407, pp. 35–50.
136. Lin, Y.C., Farr, J., Carter, K. and Fregly, B.J. (2006) Response surface optimization for joint contact model evaluation. *Journal of Applied Biomechanics* 22(2), pp. 122–130.
137. Liu, S.C., Derick, L.H. and Palek, J. (1987) Visualization of the hexagonal lattice in the erythrocyte membrane skeleton. *Journal of Cell Biology* 104(3), pp. 527–536.
138. Lloyd, C. and Chan, J. (2002) Helical microtubule arrays and spiral growth. *Plant Cell* 4(10), 2319–2324.
139. Lobo, D. and Vico, F.J. (2010) Evolutionary development of tensegrity structures. *BioSystems* 101, pp. 167–176.
140. Lord, E.A. (1991) Quasicrystals and Penrose patterns. *Current Science* 61(5), pp. 313–319.
141. Lord, E.A. and Ranganathan, S. 2001a. Sphere packing, helices and the polytope {3,3,5}. *The European Physical Journal D* 15, pp. 335–343.
142. Lord, E.A. and Ranganathan, S. (2001b) The Gummelt decagon as a 'quasi unit cell'. *Acta Crytallographica A* [Online] 57(5), 531–539. [Available at: http://met.iisc.ernet.in/~lord/webfiles/actagum.pdf [Accessed 8th March 2014].
143. Lord, E.A. (2002) Helical structures: the geometry of protein helices and nanotubes. *Structural Chemistry* 13(3–4), pp. 305–314.
144. Maina, J.N. (2007) Spectacularly robust! Tensegrity principle explains the mechanical strength of the avian lung. *Respiratory Physiology and Neurobiology* 155(1), pp. 1–10.
145. Maina, J.N., Jimoh, S.A. and Hosie, M. (2010) Implicit mechanistic role of the collagen, smooth muscle, and elastic tissue components in strengthening the air and blood capillaries of the avian lung. *Journal of Anatomy* 217(5), pp. 597–608.
146. Mammoto, T. and Ingber, D.E. (2010) Mechanical control of tissue and organ development. *Development* 137(9), pp. 1407–1420.
147. Mammoto, A., Mammoto, T. and Ingber, D.E. (2012) Mechanosensitive mechanisms in transcriptional regulation. *Journal of Cell Science* 125(13), pp. 3061–3073.
148. Mandelbrot, B.B. (1983) *The fractal geometry of nature.* Henry Holt & Co.
149. Mao, J.J. (2005) Calvarial development: cells and mechanics. *Current Opinion in Orthopaedics* 16(5), pp. 331–337.

150. Mao, Y. and Schwarzbauer, J.E. (2005) Fibronectin fibrillogenesis, a cell-mediated matrix assembly process. *Matrix Biology* 24(6), pp. 389–399.
151. Martin, R., Farjanel, J., Eichenberger, D., Colige, A., Kessler, E., Hulmes, J.S. and Giraud-Guille, M.M. (2000) Liquid crystalline ordering of procollagen as a determinant of three-dimensional extracellular matrix architecture. *Journal of Molecular Biology* 301, pp. 11–17.
152. Martin, D.C. (2012) Building tensegrity models. [Online] Available at: http://www.biotensegrity.com/modeling.php [Accessed 18th March, 2014].
153. Martin, D.C. (2014) *Biotensegrity: tensional integrity in biological structures.* Munich, Keiner Press.
154. Masi, A.T. and Hannon, J.C. 2009. Human resting muscle tone (HRMT): narrative introduction and modern concepts. *Journal of Bodywork and Movement Therapies* 12(4), pp. 320–332.
155. McFadden, J. (2013) Making the quantum leap. *The Biologist* 60(2), pp. 13–16.
156. McLeod, W.T. 1987. *The Collins dictionary and thesaurus in one volume.* William Collins & Co.
157. Moore, K.A., Polte, T., Huang, S., Shi, B., Alsberg, E., Sunday, M.E. and Ingber, D.E. (2005) Control of basement membrane remodelling and epithelial branching morphogenesis in embryonic lung by Rho and cytoskeletal tension. *Developmental Dynamics* 232(2), pp. 268–281.
158. Moore, S.M., Ellis, B., Weiss, J.A., McMahon, P.J. and Debski, R.E. (2010) The glenohumeral capsule should be evaluated as a sheet of fibrous tissue: a validated finite element model. *Annals of Biomedical Engineering* 38(1), pp. 66–76.
159. Moored, K.W., Fish, F.E., Kemp, T.H. and Bart-Smith, H. (2011) Batoid fishes: inspiration for the next generation of underwater robots. *Marine Technology Society Journal* 45(4), pp. 99–109.
160. Motro, R. (2003) *Tensegrity: structural systems for the future.* Kogan Page Science.
161. Muller, M. (1996) A novel classification of planar four-bar linkages and its application to the mechanical analysis of animal systems. *Philosophical Transactions of the Royal Society of London B* 351(1340), pp. 689–720.
162. Myers, T.W. (2014) *Anatomy trains: myofascial meridians for manual and movement therapists.* 3rd ed. Elsevier.
163. Nelson, C.M. (2009) Geometric control of tissue morphogenesis. *Biochimica et Biophysica Acta* 1793(5), pp. 903–910.
164. Nelson, C.M., Jean, R.P., Tan, J.L., Liu, W.F., Sniadecki, N.J. and Spector, A.A. (2005) Emergent patterns of growth controlled by multicellular form and mechanics. *Proceedings of the National Academy of Sciences* 102(33), pp. 11594–11599.
165. Nevin, C. (2013) Personal communication.
166. Nisoli, C., Gabor, N.M., Lammert, P.E., Maynard, J.D. and Crespi, V.H. (2010) Annealing a magnetic cactus into Phyllotaxis. *Physical Review E* [Online] 81(4), pp. 046107. [Available at: http://arxiv.org/pdf/1002.0622v1.pdf [Accessed 9th March 2014].
167. Olszta, M.J., Cheng, X., Jee, S.S., Kumar, R., Ki, Y.Y., Kaufman, M.J., Douglas, E.P. and Gower, L.B. (2007) Bone structure and formation: a new perspective. *Materials Science and Engineering R* 58(3-5), pp. 77–116.
168. Orki, O., Ayali, A., Shai, O, and Ben-Hanan, U. (2012) Modeling of caterpillar crawl using novel tensegrity structures. *Bioinspiration and Biomimetics* [Online] 7(4), pp. 04606. Available at: http://www.eng.tau.ac.il/~shai/studentlist_files/BioinspirBiomim.pdf [Accessed 9th March 2014].
169. Oschman, J. (2000) *Energy medicine: the scientific basis.* Churchill Livingstone.
170. Pabst, D.A. (2000) To bend a dolphin: convergence of force transmission designs in Cetaceans and Scombrid fishes. *American Zoologist* 40, pp. 146–155.
171. Palestini, P., Botto, L., Rivolta, I. and Miserocchi, G. (2011) Remodelling of membrane rafts expression in lung cells as an early sign of mechanotransduction-signalling in pulmonary edema. *Journal of Lipids* [Online] 2011, 695369. [Available at: http://www.hindawi.com/journals/jl/2011/695369/ [Accessed 9th March 2014].
172. Parker, K.K. and Ingber, D.E. (2007) Extracellular matrix, mechanotransduction and structural hierarchies in heart tissue engineering. *Philosophical Transactions of the Royal Society B* 362(1484), pp. 1267–1279.
173. Parker, K.K., Brock, A.L., Brangwynne, C., Mannix, R.J., Wang, N., Ostuni, E., Geisse, N.A., Adams, J.C., Whitesides, G.M. and Ingber, D.E. (2002) Directional control of lamellipodia extension by constraining cell shape and orienting cell tractional forces. *Federation of American Societies for Experimental Biology Journal* 16(10), pp. 1195–1204.
174. Parry, D.A.D., Fraser, R.D. and Squire, J.M. (2008) Fifty years of coiled-coils and α-helical bundles: a close relationship between sequence and structure. *Journal of Structural Biology* 163(3), pp. 258–269.
175. Pars, Marcelo. (2014) Tensegrity. [Online] Available at: http://www.tensegriteit.nl/ [Accessed 26th March 2014].
176. Parsons, J. and Marcer, N. (2006) *Osteopathy: models for diagnosis, treatment and practice.* Churchill Livingston, Elsevier.
177. Passerieux, E., Rossignol, R., Letellier, T. and Delage, J.P. (2007) Physical continuity of the perimysium from myofibers to tendons: involvement in lateral force transmission in skeletal muscle. *Journal of Structural Biology* 159(1), pp. 19–28.
178. Pauling, L. (1964) The architecture of molecules. *Proceedings of the National Academy of Sciences* 51(5), pp. 977–984.
179. Pauly, J.E., Rushing, J.L. and Scheving, L.E. (1967) An electromyographic study of some muscles crossing the elbow joint. *Anatomical Record* 159(1), pp. 47–54.
180. Pickett, G.T., Gross, M. and Okuyama, H. (2000) Spontaneous chirality in simple systems. *Physical Review Letters* 85(17), pp. 3652–3655.
181. Pienta, K.J. and Coffey, D.S. (1991) Cellular harmonic information transfer through a tissue tensegrity-matrix system. *Medical Hypotheses* 34(1), pp. 88–95.
182. Pompe, T., Renner, L. and Werner, C. (2005) Nanoscale features of fibronectin fibrillogenesis depend on protein-substrate interaction and cytoskeleton structure. *Biophysical Journal* 88(1), pp. 527–534.
183. Pritchard, J.H., Scott, J.H. and Girgis, F.G. (1956) The structure and development of cranial and facial sutures. *Journal of Anatomy* 90(1), pp. 73–86.
184. Pugh, A. (1976) *An introduction to tensegrity.* University of California Press.
185. Purslow, P.P. (2008) The extracellular matrix of skeletal and cardiac muscle. In: Fratzl, P. Ed. *Collagen: structure and mechanics.* Springer, pp. 325–357.
186. Qin, Z., Kreplak, L. and Buehler, M.J. (2009) Hierarchical structure controls nanomechanical properties of vimentin intermediate filaments. *PloS ONE* [Online] 4(10), e7294. [Available at: http://www.plosone.org/article/

info%3Adoi%2F10.1371%2Fjournal.pone.0007294 [Accessed 9th March 2014].

187. Ramachandrarao, P., Sinha, A. and Sanyal, D. (2000) On the fractal nature of Penrose tiling. *Current Science* 79(3), pp. 364–366.
188. Rapoport, D.L. (2011) On the fusion of physics and Klein bottle logic in biology, embryogenesis and evolution. *NeuroQuantology* 9(4), pp. 842–861.
189. Read, H.H. (1970) *Rutley's elements of mineralogy*. George Allen & Unwin.
190. Rhode-Barbarigos, L., Ali, N.B.H. and Smith, I.F.C. (2010) Designing tensegrity modules for pedestrian bridges. *Engineering Structures* 32(4), pp. 1158–1167.
191. Rieffel, J.A., Valero-Cuevas, F.J. and Lipson, H. (2010) Morphological communication: exploiting coupled dynamics in a complex mechanical structure to achieve locomotion. *Journal of the Royal Society Interface* 7(45), pp. 613–21.
192. Rifkin, B.A. and Ackerman, M.J. (2011) *Human anatomy: a visual history from the Renaissance to the digital age*. New York, Abrams.
193. Rikli, D.A., Honigmann, P., Babst, R., Cristalli, A., Morlock, M.M. and Mittllmeier, T. (2007) Intra-articular pressure measurement in the radioulnocarpal joint using a novel sensor: in vitro and in vivo results. *Journal of Hand Surgery* 32(1), pp. 67–75.
194. Robbie, D.L. (1977) Tensional forces in the human body. *Orthopaedic Review* 6(11), pp. 45–48.
195. Roos, G., Leysen, H., Van Wassenbergh, S., Herrel, A., Jacobs, P., Dierick, M., Aerts, P. and Adriaens, D. (2009) Linking morphology and motion: a test of a four-bar mechanism in seahorses. *Physiological and Biochemical Zoology* 82(1), pp. 7–19.
196. Rozario, T. and DeSimone, D.W. (2010) The extracellular matrix in development and morphogenesis: a dynamic view. *Developmental Biology* 341(1), pp. 126–140.
197. Rubin, J., Rubin, C. and Jacobs, C.R. (2006) Molecular pathways mediating mechanical signalling in bone. *Gene* 367(1), pp. 1–16.
198. Sabini, R.C. and Elkowitz, D.E. 2006. Significance of differences in patency among cranial sutures. *Journal of the American Osteopathic Association* 106(10), pp. 600–604.
199. Sadoc, J.F. and Rivier, N. 2000. Boerdijk-Coxeter helix and biological helices as quasicrystals. *Materials Science and Engineering A* 294-296(Dec), pp. 397–400.
200. Salvadori, M. 1980. *Why buildings stand up: the strength of architecture.* New York, W. W. Norton and Company, Inc.
201. Saranathan, V., Osuji, C.O., Mochrie, S.G.J., Noh, H., Narayanan, S., Sandy, A., Dufresne, E.R. and Prum, R.O. 2010. Structure, function, and self-assembly of single network gyroid ($I4_1 32$) photonic crystals in butterfly wing scales. *Proceedings of the National Academy of Sciences* 107(26), pp. 11676–11681.
202. Scarr, G. 2008. A model of the cranial vault as a tensegrity structure, and its significance to normal and abnormal cranial development. *International Journal of Osteopathic Medicine* 11, pp. 80–9.
203. Scarr, G. 2010. Simple geometry in complex organisms. *Journal of Bodywork and Movement Therapies* 14(4), pp. 424–44.
204. Scarr, G. 2011. Helical tensegrity as a structural mechanism in human anatomy. *International Journal of Osteopathic Medicine 14, pp. 24–32.*
205. Scarr, G. 2012. A consideration of the elbow as a tensegrity structure. *International Journal of Osteopathic Medicine* 15, pp. 53–65.
206. Scarr, G. (2013) Palpatory phenomena in the limbs: a proposed mechanism. *International Journal of Osteopathic Medicine* 16, pp. 114–120.
207. Scarr, G. (2014) Letter to the editor. *Journal of Manipulative and Physiological Therapeutics* 36(2), p. 141.
208. Schleip, R., Naylor, I.L., Ursu, D., Melzer, W., Zorn, A., Wilke, H., Lehmann-Horn, F. and Klingler, W. (2006) Passive muscle stiffness may be influenced by active contractility of intramuscular connective tissue. *Medical Hypotheses* 66(1), pp. 66–71.
209. Schleip, R., Findley, T.W., Chaitow, L. and Huijing, P.A. eds. (2012) *Fascia: the tensional network of the human body.* Edinburgh, Elsevier.
210. Schleip, R. and Muller, D.G. (2013) Training principle for fascial connective tissues: scientific foundation and suggested practical applications. *Journal of Bodywork and Movement Therapies* 17(1), pp. 103–115.
211. Shadwick, R. (2008) Foundations of animal hydraulics: geodesic fibres control the shape of soft bodied animals. *Journal of Experimental Biology* 211(3), pp. 289–291.
212. Silva, P.L., Fonseca, S.T. and Turvey, M.T. (2010) Is tensegrity the functional architecture of the equilibrium point hypothesis? *Motor Control* 14(3), e35–e40.
213. Skelton, R.E. and de Oliveira, M.C. (2009) *Tensegrity systems*. Springer.
214. Sorrell, G.F. and Sandstrom, G.F. (1977) *The rocks and minerals of the world.* William Collins Sons and Co.
215. Stamenovic, D. and Ingber, D.E. (2009) Tensegrity-guided self assembly: from molecules to living cells. *Soft Matter* 5, pp. 1137–1145.
216. Standring, S. ed. 2005. *Gray's Anatomy* 39th ed. Churchill Livingstone, Elsevier.
217. Stecco, A., Macchi, V., Stecco, C., Porzionato, A., Day, J.A., Delmas, V. and De Caro, R. (2009) Anatomical study of myofascial continuity in the anterior region of the upper limb. *Journal of Bodywork and Movement Therapies* 13(1), pp. 53–62.
218. Stecco, C., Gagey, O., Macchi, V., Porzionato, A., De Caro, R., Aldegheri, R. and Delmas, V. (2007a) Tendinous muscular insertions onto the deep fascia of the upper limb. First part: anatomical study. *Morphologie* 91(292), pp. 29–37.
219. Stecco, C., Gagey, O., Belloni, A., Pozzuoli, A., Porzionato, A., Macchi, V., Aldegheri, R., De Caro, R. and Delmas, V . 2007b. Anatomy of the deep fascia of the upper limb. Second part: study of innervation. *Morphologie* 91(292), pp. 38–43.
220. Stecco, L. (2004) *Fascial manipulation for musculoskeletal pain*. Padova, Piccin Nuova Libraria.
221. Stewart, I. (1998) *Life's other secret: the new mathematics of the living world.* Allen Lane, Penguin Books.
222. Still, A.T. (1899) *Philosophy of osteopathy*. Kirksville MO, (self-published).
223. SunSpiral, V., Gorospe, G., Bruce, J., Iscen, A., Korbel, G., Milam, S., Agogino, A. and Atkinson, D. (2013) Tensegrity based probes for planetary exploration: entry, descent and landing (EDL) and surface mobility analysis. *International Journal of Planetary Probes.* [Online] [Available at: http://www.sunspiral.org/vytas/cv/tensegrity_based_probes.pdf [Accessed 9th March 2014.
224. Sverdlova, N.S. and Witzel, U. (2010) Principles of determination and verification of muscle forces in the human musculoskeletal system: muscle forces to minimise ending stress. *Journal of Biomechanics* 43(3), pp. 387–396.
225. Swartz, M.A. and Hayes, M.J.D. (2007) Kinematic and dynamic analysis of a

spatial one-DOF foldable tensegrity mechanism. *Transactions of the CSME/ de la SCGM* [Online] 31(4), pp. 421–431. [Available at: http://www.tcsme.org/ Papers/Vol31/Vol31No4Paper4.pdf [Accessed 9th March 2014].
226. Tanaka, G. and Kawamura, H. (date unknown). Reference man models based on normal data from human populations. [Online] Source unknown. Available at: http://www.irpa.net/ irpa10/cdrom/00602.pdf [Accessed 9th March 2014].
227. Terayama, K., Takei, T. and Nakada, K. (1980) Joint space of the human knee and hip joint under a static load. *Engineering in Medicine* 9, pp. 67–74.
228. Termonia, Y. (1994) Molecular modeling of spider silk elasticity. *Macromolecules* 27(25), pp. 7378–7381.
229. Terrones, H., Terrones, M. and Morán-López, J.L. (2011) Curved nanomaterials. *Current Science* 81(8), pp. 1011–1029.
230. Thompson, D.W. (1961) *On growth and form.* Cambridge University Press.
231. Tietz, B.R., Carnahan, R.W., Bachmann, R.J., Quinn, R.D. and SunSpiral, V. (2013) Tetraspine: robust terrain handling on a tensegrity robot using central pattern generators. *IEEE/ASME International Conference on Advanced Intelligent Mechatronics (AIM)* [Online] Wollongong, Australia. Available at: http://ieeexplore.ieee.org/xpl/ articleDetails.jsp?arnumber=6584102 [Accessed 9th March 2014].
232. Tolić-Nørrelykke, I.M. (2008) Push-me-pull-you: how microtubules organize the cell interior. *European Biophysical Journal* 37(7), pp. 1271–1278.
233. Tschierske, C., Nürnberger, C., Ebert, H., Glettner, B., Prehm, M., Liu, F., Zeng, X.B. and Ungar, G. (2012) Complex tiling patterns in liquid crystals. *Journal of the Royal Society Interface Focus* 2(5), pp. 669–680.
234. Turrina, A., Martinez-González, M.A. and Stecco, C. (2013) The muscular force transmission system: role of the intramuscular connective tissue. *Journal of Bodywork and Movement Therapies* 17(1), pp. 95–102.
235. Turvey, M.T. (2007) Action and perception at the level of synergies. *Human Movement Science* 26(4), pp. 657–697.
236. Twarock, R. (2006) Mathematical virology: a novel approach to the structure and assembly of viruses. *Philosophical Transactions of the Royal Society A* 364(1849), pp. 3357–3373.
237. Ungar, G., Percec, V., Zeng, X. and Leowanawat, P. (2011) Liquid quasicrystals. *Israel Journal of Chemistry* 51(11–12), pp. 1206–1215.
238. Upledger, J.E. and Vredevoogd, M.F.A. (1983) *Craniosacral therapy.* Seattle, Eastland Press.
239. Usherwood, J.R., Williams, S. and Wilson, A.M. (2007) Mechanics of dog walking compared with a passive, stiff-limbed, 4-bar linkage model, and their collisional implications. *Journal of Experimental Biology* 210(3), pp. 533–540.
240. Valero-Cuevas, F.J., Yi, J.W., Brown, D., McNamara III, R.V., Paul, C. and Lipson, H. (2007) The tendon network of the fingers performs anatomical computation at a macroscopic scale. *Transactions on Biomedical Engineering* 54(6 pt. 2), pp. 1161–1166.
241. Vanag, V.K. and Epstein, I.R. (2009) Pattern formation mechanisms in reaction-diffusion systems. *International Journal of Developmental Biology* 53(5–6), pp. 673–681.
242. Van den Bogert, A.J. (2003) Exotendons for assistance of human locomotion *BioMedical Engineering Online* 2(17), pp. 1-8. Available at: http://www. biomedical-engineering-online. com/content/pdf/1475-925X-2-17.pdf [Accessed 9th March 2014].
243. Van der Wal, J. (2009) The architecture of the connective tissue in the musculoskeletal system – an often overlooked functional parameter as to proprioception in the locomotor apparatus. *International Journal of Therapeutic Massage and Bodywork* 2(4), pp. 9–23.
244. Van Workum, K. and Douglas, J.F. (2006) Symmetry, equivalence, and molecular self-assembly. *Physical Review E* [Online] 73(3), pp. 031502. Available at: http://journals. aps.org/pre/abstract/10.1103/ PhysRevE.73.031502 [Accessed 9th March 2014].
245. Verheyen, H.F. (1989) The complete set of jitterbug transformers and the analysis of their motion. *Computers and Mathematics with Applications* 17(1-3), pp. 203–250.
246. Viatchenko-Karpinski, S., Fleischmann, B.K., Liu, Q., Sauer, H. and Gryshchenko, O. 1999. Intracellular Ca^{2+} oscillations drive spontaneous contractions in cardiomyocytes during early development. *Proceedings of the National Academy of Sciences* 96(14), pp. 8259–8264.
247. Vogel, V. and Sheetz, M. (2006) Local force and geometry sensing regulate cell functions. *Nature Reviews* 7(4), pp. 265–275.
248. Waite, A.E. ed. (1894). *The hermetic and alchemical writings of Aureolus Phillippus Theophrastrus Bombast, of Hohenheim, called Paracelsus the Great;* volume 2, H3. London, James Elliot & Co. [Online] Available at: https://archive.org/details/ hermeticandalch00paragoog [Accessed 9th March 2014).
249. Walker, P.M.B. (1991) *Chambers science and technology dictionary.* Chambers.
250. Wang, N., Tytell, J.D. and Ingber, D.E. (2009) Mechanotransduction at a distance: mechanically coupling the extracellular matrix with the nucleus. *Nature Reviews* 10(1), pp. 75–82.
251. Weibel, E.R. (1991) Fractal geometry: a design principle for living organisms. *American Journal of Physiology* 261(6 pt. 1), pp. 361–369.
252. Weibel, E. R. (2009) What makes a good lung? *Swiss medical weekly* 139(27-28):375-386.
253. Weinbaum, S., Zhang, X., Han, Y., Vink, H. and Cowin, S.C. 2003. Mechanotransduction and flow across the endothelial glycocalyx. *Proceedings of the National Academy of Sciences* 100(13), pp. 7988–7995.
254. Williams, P. L. and Warwick, R. (1980) *Gray's Anatomy.* 36th ed. Churchill Livingstone.
255. Yu, J.C. (2003) Wright, R.L., Williamson, M.A., Braselton, J.P. and Abell, M.L. A fractal analysis of human cranial sutures. *The Cleft Palate-Craniofacial Journal* 40(4), pp. 409–415.
256. Zamir, M. (2001) Arterial branching within the confines of fractal L-system formalism. *Journal of General Physiology* 118(3), pp. 267–276.
257. Zanotti, G. and Guerra, C. (2003) Is tensegrity a unifying concept of protein folds? *The Federation of European Biochemical Societies Letters* 534(1-3), pp. 7–10.
258. Zhu, Q., Vera, C., Asaro, R.J., Sche, P. and Sung, L.A. (2007) A hybrid model for erythrocyte membrane: a single unit of protein network coupled with lipid bilayer. *Biophysical Journal* 93(2), pp. 386–400.

A

B

C

D

E

F

I

U

V

W